T0164918

GLORY OF NOTRUMP

by

John Sheridan Thomas

Order this book online at www.trafford.com
or email orders@trafford.com

Most Trafford titles are also available at major online book retailers.

Printed in the United States of America.

ISBN: 978-1-4269-7634-6 (sc)
ISBN: 978-1-4269-7635-3 (e)

Library of Congress Control Number: 2011914124

Trafford rev. 09/27/2011

www.trafford.com

North America & international
toll-free: 1 888 232 4444 (USA & Canada)
phone: 250 383 6864 ♦ fax: 812 355 4082

FOREWARD

Some 60 years ago Charles Goren developed an integrated bidding system that quickly became the standard for virtually every bridge player in North America. What made it especially unique was his point-count method used to evaluate bridge hands. The simplicity and utility of Goren's system led it to be adopted throughout North America, and so it soon became known as Standard American.

As one would expect, there have been huge advances in bidding technology (bidding rules if you prefer the mundane) since Goren's superb achievement. Naturally most of the attention has been directed toward duplicate bridge. This is not surprising since competitive players have the most at stake in getting it right. And by-and-large they have done so.

Over the years it became apparent to me that both social and tournament players under-rate notrump by a country mile. Consequently I developed a seminar called the **Glory of Notrump** aimed at enabling students to find and play good notrump contracts which otherwise were being played in inferior suit contracts. It was a natural evolution from that seminar to this book.

You will find explicit recommendations herein. They are supported by extensive research and statistical analyses that have not been applied heretofore to contract bridge.

Glory of Notrump is replete with examples of experts bidding and playing notrump contracts in world championship venues. Assuredly your bridge game will be much better for having read it.

*   *   *   *

Artwork

Each chapter is introduced with a sketch intended more or less to mimic the topic of the chapter. These sketches were created and drawn by my brother Augustus O. Thomas, II (Bud). Bud is the author of THOMAS THOMAS DESCENDANTS, published in 2008. Before he became a writer, he was an electrical engineer; and long before that when a young boy, he was a cartoon artist of wide repute. It seemed fitting to call upon his long-dormant artistic skills to dress up this book; besides the price was right.

* * * *

Dedication

This book is dedicated to Marian Wolfe Thomas, my beloved wife and numero uno bridge partner for 60 years and counting,

CONTENTS

INTRODUCTION

Just ten points! In contract bridge, where games and slams score 300 points and up, scoring ten points more or less is relatively miniscule. I am referring of course to the incremental value of the first trick in notrump contracts, assuming you make your contract. In duplicate bridge, ten points produce match-point gains as valuable as any game or slam. Alternatively in rubber bridge ten points is a mere ten points, except that you need only nine tricks instead of ten or eleven in suits.

Notrump

Have you noticed how some bridge players avoid notrump contracts in favor of suits? These are substantially different games - suits vs. notrump. In suits you need not worry much about those short, weak ones, and often it is wise to concede a side loser or two in order to obtain a ruff.

As we all know too well, playing in notrump requires different tactics. Easily the most important task is to establish slow winners (especially long suits if you are lucky enough to have any) before the defenders drive out your stoppers in their best suit. It is all about timing. It is easy to play top cards where you have them. But when you do not have enough quick winners, you must find a way to boost lesser-ranked cards into winners; and to accomplish this before it is too late. If you need nine tricks and can see a way to get them, it is of no avail if the defense manages to get five first.

Occasionally notrump contracts require extraordinary skill to succeed. Experts have those unique faculties of tracking cards that have been played or not, and of visualizing the shape and substance of defenders' hands as play proceeds. Fortunately for the rest of us, most good notrump contracts can be made with more limited faculties, provided we have acquired a sufficient, but indefinable, amount of experience.

When you play difficult contracts, quite frequently success depends on the location of a key card or maybe two or three; and at times you have multiple choices as to how to proceed. So, in addition to experience, you need some basic understanding of probabilities (make that odds) to be able to select the most profitable line of play.

Declarers get to see all of their assets while defenders must search for hints to find good lines of play; and quite often they do not succeed. This is especially true in notrump where time and again the defense, to prevail, must quickly find declarer's weakness. Yet so often they are unable to do this. In fact it happens with such regularity that it is quite tempting to bid notrump games with less than 26 points. Perhaps with deals as light as - - . Oops, I am getting ahead of the story.

Threads

There are three major threads to this story. One is the notion of shape: how the shape of your hand impacts on the number of tricks you can take. This is an obvious relationship when we are dealing with suit contracts. Trumps are superior cards - the more you have, the more tricks you take; and short suits enable more tricks by ruffing.

The roll shape plays in notrump contracts is more subtle and generally less understood by players at all levels of the game. Long suits produce slow winners in notrump contracts just as they do in suit contracts. Of course you need to get these slow winners before the defenders get theirs. The surprise is how large a role slow tricks play in notrump. Length begins at four, and it is notable how often fourth and fifth spots can win tricks for you.

Another thread to this story is that standard bidding does not get us into good notrump contracts nearly as often as it should. Often notrump is bypassed in favor of a suit contract. And even when notrump is given its proper respect, we tend to under-rate one or another of the partnership hands and thus do not bid game often enough.

This suggests a lack of precision or lack of confidence in partnership bidding methods. In uncertainty, staying with the opposition is a conservative posture. However if you knew with some certainty the odds of making three notrump were say 40%, you would not take that chance no matter whether the opponents were likely to go for it or not; because in the long run, it is not profitable to go against mathematical odds.

Admittedly if you are behind in the fourth quarter, there may be no way to catch up except to take chances; and that could mean going for big swings where thin game tries and marginal penalty doubles may be the only means to make up a large deficit. My mathematical view of such is this: going for say 40% games in the fourth quarter can improve your chance of winning if the alternative is near-certain defeat.

In close calls, there are qualitative factors to consider when making the decision of game or no. Actually any hand characteristic that we do not assign point values is thrown into the category of 'qualitative factors'. This is not a matter of preference; we do it because of an inability to assign hard point values. These factors include suit quality, intermediates, aces/kings, quacks, worthless doubletons, and so on. One factor, suit quality, can be assigned point values; and it is one of several ways to increase the accuracy of notrump game decisions.

The third thread is to become more aggressive, in a disciplined way, when bidding notrump contracts. After reading this book you should find yourself playing more difficult contracts; surprisingly some will be easy to make; others will need exquisite judgment; still others will depend on the good graces of your opponents; and of course a few will be quite impossible to bring home, even where the opponents inadvertently help.

Competitive Edge

Glory is about more than these several threads. It is about ways to achieve a competitive edge whenever and wherever notrump is appropriate. You can gain this edge by applying more aggressive practices incrementally as you integrate them into your partnership game.

Surely many readers already practice some of the winning practices presented herein. Even so, no matter what level of bridge skills you possess, from intermediate to expert, you will gain a better understanding of the significance of various hand distributions and how they contribute to trick-taking in notrump. Without tongue-in-cheek, I can promise your bridge game will improve in one manner or another by carefully reading this text.

Experts at Play

Most of the deals presented are from world tournament competition. The nature of notrump, and indeed bridge itself, with billions of unique deals, make it difficult to devise bidding and play rules to accommodate more than a fraction of the possible combinations and permutations. This characteristic places a premium on long-standing partnerships and extensive experience at the table. Studying how these deals were played will most assuredly accelerate development of your bridge expertise.

The World Bridge Federation publishes results of world championship tournaments, including the Bermuda Bowl, the Venice Cup and the World Olympiad. These publications contain detailed hand records of every deal played in the semi-finals and the finals. They are a key source of how world-class experts bid and play the game. Most of these contests are teams-of-four and scoring is in International Matchpoints, or IMPs for short. Not all of the lessons learned from study of IMP contests are directly transferable to match-point duplicate. An obvious example involves overtricks. Overtricks can be crucial in match-point bridge where 30 points can mean the difference between a top board and an average board. On the other hand 30 points gained from an over-trick translates into just one IMP in a team contest where it takes hundreds of IMPs to win.

There are many world championship deals presented and examined herein. This is the world of expert bridge, and their play is instructive for the rest of us. Included are deals played by renowned bridge players such as Robert Hammon, Robert Wolff, Eric Rodwell, Jeffery Meckstroth, Lew Stansby, Lynn Deas, and Stasha Cohen. These are just a few of many world-class players featured.

Alert

You will notice throughout the text that there are deals from both pairs and teams-of-four contests. Both formats are played extensively in tournaments. However team contests have dominated in World Championship venues, and a majority of the deals presented are from these team contests. Please do not assume that this book is intended primarily for team bridge players. Deals from team contests were selected because of the availability of detailed play records published by the World Bridge Federation. The fundamental skills for success in any form of contract bridge are the same even though the tactics differ in some respects.

For those who are not familiar with team contests, a brief description is presented in Appendix C.

1
MYSTIQUE

World-class bridge players take tricks and make seemingly impossible contracts with a frequency that is beyond normal, sometimes so far beyond as to be a bit scary. Some of it is luck of course; but there is no reason for experts to have more luck then the rest of us. After all, we hold the same hands, play the same deals.

Surely they, the experts, have developed super card sense and tons of experience; and of course they play with some pretty good partners. Is that the whole of it or is there something else? Can they really "see through the backs of the cards" as Audrey Grant might suggest?

Here are several deals to ponder; an eclectic collection of world tournament play; deals where declarers manufacture, or seem to, tricks beyond what skill can explain. Maybe they intimidate defenders into making mistakes. Or maybe it is a mystique, an enigmatic persona that enables him or her to find obscure winning lines of play.

* * * *

BERMUDA BOWL - 1999. Early in the final team contest between USA1 and Brazil, Brazil was leading 23-13. The intrepid Mechstroth/ Rodwell pair (MECKWELL) held these North-South cards.

Board 9
Dealer North
East-West Vulnerable

NORTH
♠ K Q 6
♥ J 10 2
♦ K Q 8 7 5 2
♣ 6

WEST
♠ J 8 2
♥ A 4
♦ A 4
♣ K J 9 7 4 3

EAST
♠ 10 7 5 4
♥ K Q 9 7 5
♦ 10 3
♣ 8 2

SOUTH
♠ A 9 3
♥ 8 6 3
♦ J 9 6
♣ A Q 10 5

West	North	East	South
	1♦	Pass	2♣
Pass	2♦	Pass	2♥ (Relay)
Pass	2NT	Pass	3NT ///

Rodwell (South) was not shy about forcing to game with his eleven-point hand. Notrump lite indeed: 22 high-card points between them. However Meckstroth's six diamonds looked like they might be the salvation.

In fact there were plenty of tricks if Meckstroth could extract the ♦A before losing a bunch of hearts. East selected the obvious good heart lead. Unfortunately East chose the ♥K, overtaken with the ♥A by West, who immediately returned his remaining heart. Of course East hoped to drive out Meckstroth's sole heart stopper early then run hearts whenever partner got into the lead. However this first trick created a heart stopper for

Meckstroth and prevented West from returning a heart when he won his ♦A. Meckstroth was able to concede a diamond without fear of another heart return.

Elsewhere Brazil also reached 3NT but from the other side, putting West on lead. West began with a low spade; declarer winning and immediately driving out the ♦A. At the second chance to find hearts, West concluded that a switch was called for, but selected clubs. Ah, too little, too late.

Both teams had aggressively pressed this very lite deal to improbable games, no doubt counting on diamonds to come through, or maybe just wishing it so. In any event, both sides made the improbable game - serendipity, perhaps with a touch of mystique.

* * * *

WORLD TEAM OLYMPIAD - 1996. One of the semi-finals found Indonesia seated against Denmark. Denmark bid this deal to 3NT even though both hands were highly skewed toward the minors.

Board 95 (Rotated)
Dealer West
East-West Vulnerable

	NORTH	
	♠ 6 3	
	♥ J 8	
	♦ A 7 3 2	
	♣ A Q 10 9 5	
WEST		**EAST**
♠ J 10 7 5 4		♠ A 9 8 2
♥ K 7 6 3		♥ Q 10 9 4 2
♦ J 8		♦ K
♣ K 2		♣ J 8 3
	SOUTH	
	♠ K Q	
	♥ A 5	
	♦ Q 10 9 6 5 4	
	♣ 7 6 4	

West	North	East	South
Pass	1♣*	Pass	1♠*
Pass	2♣	Pass	2NT
Pass	3♦	Pass	3NT ///

* Conventional

North-South bid the notrump game in a "conventional" manner. West led a fourth spade to partner's ♠A, and a spade return eliminated South's (Christiansen) sole spade stopper. One more loser and the game would be up. Christiansen played the ♦10 - ♦8 - dummy ♦A, dropping the singleton ♦K! Christiansen now had the contract secure and did not need the club finesse - taking six diamonds and one in each of the other suits.

This feels like pure luck; but reflect on the diamond play. By leading the ♦10 to the ♦A, Christiansen tried to entice an honor from West. Failing to do so, he had no choice but to play the ace and hope East had the bare ♦K.

This deal was played at three other tables. None tried notrump and none made game.

* * * *

SENIOR BOWL - 2007. The Seniors Bowl was a recent (2001) addition to World Bridge Championships. USA teams won every one of the first three championships. In this, the fourth Seniors, USA1 was well ahead of France in the quarterfinals, and they won the 96-board match 215 to 175.

Board 87 (Rotated)
Dealer East
All Vulnerable

NORTH
♠ A
♥ K Q 8
♦ A K Q 8 7
♣ J 10 9 5

WEST
♠ K J 7 4 3
♥ A 10 5 2
♦ J 9
♣ Q 7

EAST
♠ 9 6 5
♥ 9 7 4
♦ 5 3 2
♣ A 8 6 4

SOUTH
♠ Q 10 8 2
♥ J 6 3
♦ 10 6 4
♣ K 3 2

West	**North**	**East**	**South**
		Pass	Pass
1♠	Dbl	Pass	1NT
Pass	3NT///		

Sitting North-South were Morse and Wolff, USA1, with Wolff the declarer in 3NT. West led his fourth spade to dummy's ♠A. Wolff cashed one high diamond, West false-carding with the ♦J. Next Wolff lead the ♣J to East's ♣A, who then returned a spade, his partner's opening suit. West took two spades, and led another spade to Wolff. Wolff ran diamonds; then led to the ♣K, West's ♣Q falling under it. Another club gave Wolff his ninth trick. The defenders managed to take just two spades and two aces.

In the other room North-South (France) played the same contract with the same spade opening lead, and was set two tricks. The second time clubs were led (♣10 from dummy), South played low in his hand and lost the trick to West's bare queen. At the other table Wolff somehow worked it out that East was not likely to hold both the queen and ace of clubs.

* * * *

SENIOR BOWL - 2007. Having defeated France in the quarterfinal, USA1 went on to play USA2 in the semifinal. Here was a classic: playing West were Wolff for team USA1 and Stansby for USA2; both declaring 3NT. Collectively Wolff and Stansby had won 17 World Championships!

Board 8 (Rotated) Dealer South None Vulnerable	**NORTH** ♠ 8 6 2 ♥ K J ♦ A Q 10 4 ♣ Q 9 7 6	
WEST ♠ K 10 7 4 3 ♥ 6 4 3 ♦ 9 8 ♣ A K 4		**EAST** ♠ 9 5 ♥ 10 9 8 7 ♦ K J 6 5 2 ♣ J 10
	SOUTH ♠ A Q J ♥ A Q 5 2 ♦ 7 3 ♣ 8 5 3 2	

West	North	East	South
		1♣	
1♠	2♠	Pass	2NT
Pass	3NT///		

Both teams arrived at 3NT via the same bidding. West led his fourth spade to declarer's ♠J. Stansby immediately played clubs - 3♣, ♣4, ♣Q, ♣10; then another club from dummy - ♣6, ♣J, ♣2, to West's ♣K. West switched to a heart, won in dummy. Stansby won his second heart in dummy and led another club to West's ♣A. A diamond shift was too late; Stansby had his contract - two spades, four hearts, one diamond and two clubs.

In the other room, Wolff (USA1) played clubs in the same manner as Stansby, also making his contract for a tie board.

Would you have played clubs the way Wolff and Stansby had? It cannot be too obvious because in the other semifinal contest between Brazil and Indonesia, also in 3NT, neither played clubs as had Stansby and Wolff, and both paid the price; Indonesia down two, Brazil down three. Those declarers allowed the defense time to find their rightful diamond and spade tricks.

USA2 won this match and went on to defeat Indonesia 205 to 127 for the 2007 championship.

*   *   *   *

VENICE CUP - 2000. In the round robin eliminations, Great Britain's Dhondy backed China into a corner on board 20 with no option but to hand her the game-winning trick.

Board 20 (Rotated)
Dealer South
Both Vulnerable

WEST
♠ K 9 5 4
♥ 2
♦ A 10 6 5
♣ K 5 4 2

WEST
♠ A 10
♥ 10 8 5 3
♦ K 8 2
♣ Q J 9 8

EAST
♠ Q 8 6 3 2
♥ K 6 4
♦ 7 3
♣ 7 6 3

SOUTH
♠ J 7
♥ A Q J 9 7
♦ Q J 9 4
♣ A 10

West	North	East	South
			1♥
Pass	1♠	Pass	2♦
Pass	3♦	Pass	3NT ///

West led the ♣Q to declarer's ♣A (Dhondy). Fortunately both the ♦K and the ♥K were onside. Even so, Dhondy could finesse hearts only once and thus take no more than eight tricks without help from the defense.

Dhondy played the ♦Q to the king and ace. Then she finessed to the ♥Q. Next she cashed three diamonds and the ♣K, exiting dummy with a club. West cleared clubs and led the ♠A and ♠10. Dhondy ducked the ♠10 to East and had the remaining tricks whatever East returned.

* * * *

We reviewed six deals, none exotic, nevertheless involving faultless declarer play. Such superior performances in World Championships are legendary. You may sense a mystical aura around the some of them. Some can be explained by technique; and we can identify and describe these. However there still remains something else - a mystique!

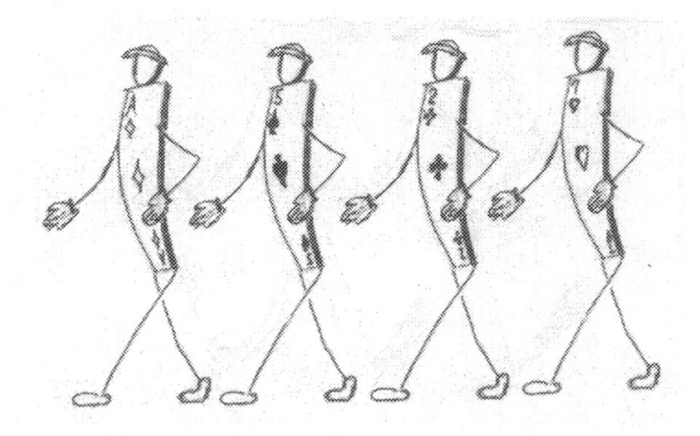

2

FLATS

Flats: hands with a distribution pattern of 4-3-3-3 (in any order). Technically these hands are not precisely flat; but they are flatter than any other suit distribution you can get from a standard deck of cards.

Flats are not especially beneficial; in fact mostly they are detrimental. Hand evaluations recognize this, but only obliquely by assigning values to shortness or extra suit length.

Flats and the Golden Fit

Suppose South has a hand qualifying to open 1NT (15 to 17), with a distribution of 5-3-3-2, and the five is a major. Simultaneously North's hand is flat (4-3-3-3 in any suit order), thus producing a major suit 5-3 golden fit. Additionally there is a 25% possibility of a 5-4 fit in instances where North's four-card suit happens to be the same as South's five-card suit. For the moment, our focus is on 5-3 fits.

I was conditioned from early novice days onward to the idea that whenever a golden fit is present in a major, the contract should be played in that suit. Basically it is a good idea and generally true; but there are many exceptions.

Let's consider three relevant deals covering the spectrum of usual and common results. They include one where the major suit is better, one where notrump wins; and a third which produces the same number of tricks either way.

*   *   *   *

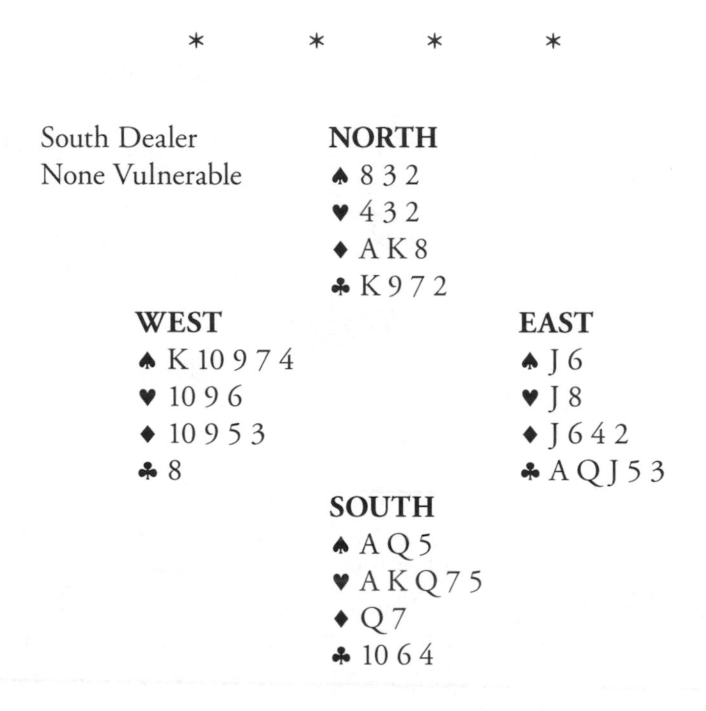

At Table 2, South opens a heart. North has support and 10 points, so they easily continue to a 4♥ game. West leads the singleton club; declarer plays dummy's king; and East takes the trick and continues clubs. The fourth club promotes a trump trick for West. I suppose declarer should play the nine on the first trick, but in the end it doesn't matter as declarer has only nine tricks, that is unless the defense gives him the ♣K.

The board is passed to Table 1, where North-South bid to 3NT. West leads his fourth spade into declarer's ♥A -♥Q, giving him ten tricks off the top; winning the board and a difference of 480 points in favor of notrump. As fate would have it, West's singleton club is a bad break for the heart bidders, while notrump bidders get a break from the spade lead.

*　*　*　*

South Dealer
Both Vulnerable

NORTH
♠ 8 4 3
♥ Q J 10
♦ K 7 6 4
♣ K 10 6

WEST
♠ 10 9
♥ A 7 4 3
♦ Q 10 9 2
♣ A 5 2

EAST
♠ J 5 2
♥ 9 8 5
♦ A 8 5
♣ J 9 8 7

SOUTH
♠ A K Q 7 6
♥ K 6 2
♦ J 3
♣ Q 4 3

At table 2, South holds a strong 15-point hand and finds partner with spade support. These hands are a little shy of game so let's presume South is playing 2♠. West leads the ♦10. East plays low, holding his ace behind dummy king. South takes the trick with the ♦J. South easily takes eight more tricks (five spades, two hearts, one club) to make an overtrick, +140.

The board moves to Table 1, where the bidding goes 1NT - - 2NT and South is declarer. West leads the ♦2 to the ♦6, ♦A, ♦3. East returns the ♦8, ♦J, ♦Q, and ♦K in dummy. Now South can run spades but sooner or later West gets in with an ace and takes two diamonds and another ace - plus 120 points but no cigar! By favoring notrump, North-South loses the board by 20 points. The key factor is declarer's short diamond suit. In a spade contract, this shortness prevents the defense from obtaining another diamond trick.

*　*　*　*

South Dealer
None Vulnerable

NORTH
♠ Q 8 7 3
♥ Q 7 3
♦ 6 4 3
♣ A 10 8

WEST
♠ K J 6 2
♥ 10 6 2
♦ Q 8
♣ Q 9 7 5

EAST
♠ 10 9 4
♥ K J
♦ K J 10 9 5
♣ 6 3 2

SOUTH
♠ A 5
♥ A 9 8 5 4
♦ A 7 2
♣ K J 4

You (South) open 1♥, partner raises to 2♥ and you pass out. West leads the ♣5. Conveniently the club lead gives you three club tricks. I presume you would promptly play the ♥A and notice East's play of the ♥J. If this is not a false card, East will drop the king next or else West began with four hearts. Either way, you should lead the ♥9 and finesse West for the ♥10. In this manner you lose only one heart. If East does not switch to diamonds immediately, you can make the ♠Q good and take a total of 10 tricks.

Surely some would bid and play this deal in notrump rather than hearts, but whatever the contract, the same 10 tricks can be taken playing in either strain. In notrump, even if East switches to diamonds, South can still take 10 tricks by holding up on the first diamond lead. When the skill and luck of the contestants are equal, notrump wins the deal - - but by a mere 10 points. Perhaps you would bid these North-South hands to game, but that's a story yet to be told.

*　　*　　*　　*

The foregoing three deals represent a normal spectrum of results with 5-3 majors opposite flat hands. To briefly summarize them:

> The first deal was lost by the suit contractor because one defender had a singleton and was able to ruff early. These occasions are infrequent; when the offensive hands are balanced or nearly so, the defensive hands tend to be balanced too.
>
> The second deal was won by the suit contractor by one trick because the trump suit protected him from an additional diamond loss. This result was mainly due to an unfavorable placement of diamond honors, a random event that I would expect to find somewhat less than one-half of the time.
>
> The third deal provided no favors to either party; consequently the same number of tricks was available in notrump as in the suit contract. This is a common outcome where the dummy is flat, providing no ruffing tricks. In deals where the trick-taking potential is equal, notrump is the winner, most often by 10 points; but when exactly nine tricks can be taken, notrump wins by a game margin.

Notrump is the favored contract in match-point bridge where a ten-point scoring advantage so often is the difference between an above average and a below average board. Keep in mind the balanced profiles of these deals.

Golden fits are not just 5-3 majors; they include 4-4 majors too; and the same preference for notrump extends to them whenever and wherever one hand is flat. A singular notice: these 4-4 fits occur more often than 5-3 fits.

* * * *

1994 - World Open Pairs Finals. In this deal we have the all-to-common choice of a 4-4 major or notrump with one hand flat.

Board 10 (Rotated)
South Dealer
Both Vulnerable

NORTH
♠ Q 4 2
♥ J 10 6 3
♦ Q
♣ A K 7 3 2

WEST
♠ A 10 7 6
♥ K 2
♦ 7 5 4 3 2
♣ J 8

EAST
♠ 9 5 3
♥ A 7 4
♦ A 10 9 6
♣ 10 9 6

SOUTH
♠ K J 8
♥ Q 9 8 5
♦ K J 8
♣ Q 5 4

West	**North**	**East**	**South**
			Pass
Pass	1♣	Pass	1♥
Pass	2♥	Pass	3♣
Pass	3♥///		

North-South (Swedes) had 24 high-card points, a good club suit, and a 4-4 heart suit; but South had the dreaded 3-4-3-3 flat hand. At fourteen other tables the North-South pairs went on to 4♥, and paid the price. Here the Swedes stopped short in 3♥, and they took their nine ticks, losing three aces and the trump king.

Elsewhere nine pairs bid and made 3NT for top scores. In theory the notrump contract could have been defeated if East held up on the first diamond lead, but none did; it was too tempting to overtake the dummy singleton queen and return partner's lead. As anticipated, the notrump heretics walloped the major traditionalists - all because South's hand was flat.

*   *   *   *

There are many deals favoring notrump over majors; actually more than many, a multitude:

> **Notrump triumphs over golden-fit majors by a margin of about 2:1 - that is two out of every three deals - where the supporting hand is flat.**

The most common situations are those where both majors and notrump have the same trick-taking potential because the flat hand is unable to produce ruffing tricks.

To Stayman Or Not?

Suppose partner opens 1NT with 4-3-3-3, 4-4-3-2, or 5-3-3-2 (five card major). Now further suppose you have a flat hand (4-3-3-3) and your four-card suit is a major. We now know a 5-3 suit is inferior to notrump when the supporting hand is flat. It is also established, I trust, that a 4-4 fit in these otherwise same configurations is also less alluring than notrump.

WEST	**EAST**
♠ K Q 6 5	♠ J 9 3 2
♥ 8 6 4	♥ A Q J 7
♦ A 5 4	♦ Q 3 2
♣ 10 8 7	♣ A K

East opens 1NT. West has nine high-card points, surely sufficient to invite game. Typically we employ Stayman to investigate a 4-4 spade fit. But not here. West raises to 2NT. Good call - bypassing Stayman with this flat hand. Holding the maximum 17 high-card points, East continues to 3NT.

This deal should produce nine tricks nearly every time. There are two stoppers in the minors and quite possibly three tricks in each major. But making the contract is not quite the point. Rather the point is a suit contract does not offer any more tricks than notrump. In this instance, there is the highly satisfying consequence of making game in notrump while others are going down in 4♠.

*   *   *   *

SENIOR BOWL - 2007. Steering flat hands into notrump has broader applications than simply responding to notrump openers. The objective - to play in notrump rather than a 5-3 or 4-4 major when either hand is flat - does not differ when the bidding follows a different path, a point not lost on this Argentina team playing against Indonesia.

Board 22
East Dealer
East-West Vulnerable

NORTH
♠ A 9 7 4
♥ A Q 6
♦ A K 4
♣ Q 9 4

WEST
♠ K 2
♥ K 10 9 7 5
♦ 10 9 8 6
♣ K 6

EAST
♠ 10 8 5
♥ 3
♦ J 5 3 2
♣ 10 8 7 5 2

SOUTH
♠ Q J 6 3
♥ J 8 4 2
♦ Q 7
♣ A J 3

Beginning with a strong club opener, Indonesia bid to a spade game with North (Lasut) declarer. East led his singleton heart and West played the nine, hoping to deceive declarer. Later when West won the ♠K, he promptly returned hearts for partner to ruff. An alert declarer played the ♥Q which East promptly ruffed; however the queen in this instance was a loser on loser. Lasut had saved his ace and the contract.

In the other room there was no attempt to find a 4-4 fit; Argentina bid directly to 3NT and had an easy ten tricks. The deal was a push in IMPs, but it would have been a clear victory, 10 large ones, in match-point.

*   *   *   *

Super Fits

Ah yes - super fits! We have this other possibility when partner opens 1NT with a five-card major and your hand is flat. This is the possibility that responder's four-card major matches opener's five-card major. Do we miss better suit contracts by focusing on notrump where a 5-4 major exists?

South Dealer
East.-West Vulnerable

NORTH
♠ 10 9 6 2
♥ A Q 5
♦ A 10 9
♣ 10 7 3

SOUTH
♠ A J 8 7 3
♥ K 9 8
♦ K 8
♣ A 9 2

You (South as usual) open 1NT and partner employs Stayman to find a 4-4 fit in spades, then continues to a spade game. (If you are wondering why open 1NT with a five-card major, hold that thought; it will be given a thorough airing anon.)

You are South and 1NT. Then partner puts you into a four spade contract via Stayman. The full deal is:

South Dealer
East.-West Vulnerable

NORTH
♠ 10 9 6 2
♥ A Q 5
♦ A 10 9
♣ 10 7 3

WEST
♠ 5
♥ 7 6 4 3 2
♦ Q 3 2
♣ J 8 6 4

EAST
♠ K Q 4
♥ J 10
♦ J 7 6 5 4
♣ K Q 5

SOUTH
♠ A J 8 7 3
♥ K 9 8
♦ K 8
♣ A 9 2

Lacking a way to attack, West leads the ♣4 to East's ♣K and your ♣A. You are facing two club losers so you will have to keep your trump losses to just one. Of course you go to dummy and lead spades until East plays the king or queen, which you capture with your ace. Well done; you are home free with four spades, three hearts, two diamonds and a club.

Alternatively suppose after South opens 1NT, North jumps directly to 3NT. West's hearts are hopeless so he leads the ♣4. Declarer counts winners, finding one club, two diamonds, three hearts, and hopefully four spades, more than enough to make the contract. However South must be attentive and duck the first two club tricks. Also he needs some luck that East has both spade honors; because if West gets into the lead, he will take a fourth club, keeping the declarer to nine tricks, making the contract but losing the board by 20 points.

Taking ten tricks in both spades and notrump favors notrump by 10 points, but here notrump requires some luck and more alert play than does the suit contract.

* * * *

An analysis of random deals where the partnership has a 5-4 major fit and both hands are otherwise balanced shows an overwhelming margin of nearly 60% to 40% in favor of notrump. This comparative win-loss rate is critical in match-point contests.

The Pro Line

In match-point duplicate the odds of winning deals where one hand is flat enormously favor notrump over majors. It does not matter if the major is 4-4, 5-3 or even 5-4 so long as one hand is balanced and the other flat. Favoring notrump is the only way you can be sure your progressive competitors do not get that 10-point edge.

You must expect occasions where you lack an essential stopper; it comes with the territory. Fortunately the defense with notable regularity does not exploit this weakness, but when it does - - well those are the deals you wish you had been in the suit contract. You must be prepared to take the bitter along with the sweet.

Lest you are tempted to continue to favor majors, you should be mindful that one of every ten hands is flat; and when you add near-flat hands (4-4-3-2) they occur nearly one of every three deals.

Please note this exception. When playing rubber or team bridge the statistics favor 5-4 fits in the major. Here the transfer mechanism is a handy tool when responder has the five-card major. However when the opener has it and opens 1NT, responder with a flat hand will unknowingly opt for the notrump contract. Ah well, it is an imperfect world after all.

The Bottom Line

Shapes trump flats . . . even in notrump.

3

SHADOWS

Bridge defenders dwell in shadows where visibility is so poor they often stumble around while the offense effortlessly races to its goal. Alas poor defenders; this is their natural condition. They have some tools, but they are axes where scalpels are needed. Such crude tools can improve their lot, but only marginally. I do not mean to belittle their efforts; they do what they can with limited resources and less information, employing studied efforts to signal partner; making lead-directing bids; and less often, but not reliably, listening closely to over-zealous opponents anxious to describe their hands to a tee. Yes, many bridge pairs are quite proud of their ability to describe their assets to partner precisely, but in doing so, every tidbit informs and enlightens their opponents equally.

Other than what little they can glean from the bidding, defenders are at such disadvantage that they frequently bestow unearned tricks upon declarers. By and large this condition is pervasive, and it provides declarers with abundant opportunities to win marginal tricks. Keeping the defense in shadows gives declarers the winning edge in a surprising number of deals.

* * * *

BERMUDA BOWL - 1991. There are times, though infrequent, that a USA team does not make it to the finals. This year it was Iceland vs. Poland. Nearing the end, Iceland was ahead by 56 IMPs, but Poland was not conceding, not about to fold.

Board 155 (Rotated)
West Dealer
None Vulnerable

NORTH
♠ K 8 6 4 2
♥ Q J 5 3
♦ J 2
♣ Q 6

WEST
♠ A 7 5
♥ 10 9 7 6
♦ Q 6 5 3
♣ J 7

EAST
♠ 9 3
♥ K 4
♦ 10 9 8 7
♣ A K 8 5 4

SOUTH
♠ Q J 10
♥ A 8 2
♦ A K 4
♣ 10 9 3 2

West	North	East	South
Pass	1♣ *	Pass	3NT ///

* 7 to 12 HCP, majors

South (Poland's Balicki) was not going to miss game if it was at all possible, and he wasted no time getting there. West began with the ♥10 to the queen, king, and ace. Balicki could count four spades, two hearts and two diamonds, for eight tricks. The ninth trick would have to come some how from hearts or clubs.

Balicki proceeded to establish spades, West holding up his ♠A until the third round. It wasn't clear to West where to attack so he continued hearts, leading the ♥9, taken by the ♥J in dummy. This of course established declarer's ♥8 for his ninth trick and game.

Note that South, holding a flat hand, went directly to the notrump game in spite of partner's interest in the majors. Also it is worth noting that North-South had 23 high-card points, plus a good five-card suit, plus two tens, one nine and two eights. All together, it was not a bad game try.

*   *   *   *

VENICE CUP - 1993. In the round-robin, sixteen teams were competing for four places in the semifinals. This deal featured New Zealand verses India.

Board 19
South Dealer
East-West Vulnerable

NORTH
♠ A J 7
♥ K 9 8 7 4
♦ A K 6
♣ 6 4

SOUTH
♠ 8 2
♥ 10 6
♦ Q 4 3
♣ A J 9 8 5 3

West	North	East	South
			Pass
Pass	1♣*	Pass	1♦
Pass	1NT	Pass	3♣
Pass	3NT ///		

* Conventional

The defense gained little insight from this bidding which began with a strong club opener. North (Wilkinson) was declarer. East led a low heart to the ♥J and ♥K.

The full deal was:

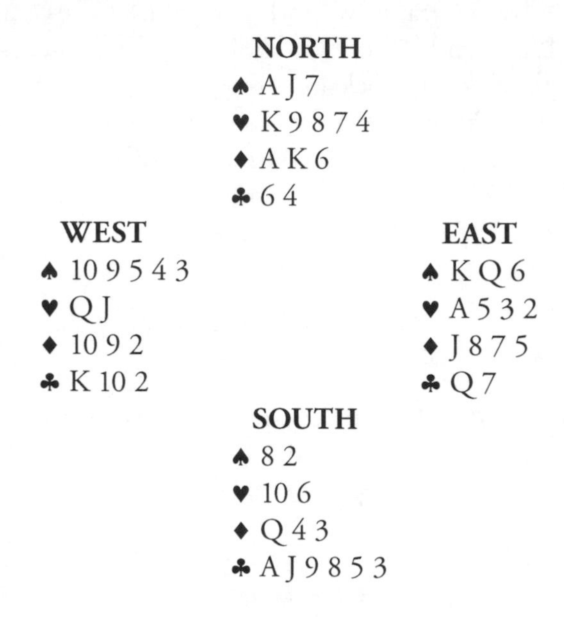
NORTH
♠ A J 7
♥ K 9 8 7 4
♦ A K 6
♣ 6 4

WEST
♠ 10 9 5 4 3
♥ Q J
♦ 10 9 2
♣ K 10 2

EAST
♠ K Q 6
♥ A 5 3 2
♦ J 8 7 5
♣ Q 7

SOUTH
♠ 8 2
♥ 10 6
♦ Q 4 3
♣ A J 9 8 5 3

Wilkinson's only chance was to win four of six clubs, so she began with a low club and ducked when East played the ♣Q. East led a second heart, taken by West's ♥J, who then switched to a spade. This looked like the defense was going to take two clubs, two hearts and two spades to set the contract. However Wilkinson ducked the spade to East's ♠Q. Unable to see through the fog, East switched to a diamond which gave the tempo back to Wilkinson. She now had time to drive out the remaining ♣K in West's hand. West had no hearts left so East had to swallow the ♥A. Wilkinson made her game with twenty-two high-card points!

* * * *

WORLD OPEN PAIRS - 1994. Seventy-five pairs competed in this match-point championship, including 19 from the United States. This deal features Americans Katz and Weinstein in the finals against Poland's Lesniewski and Szymanowski.

Board 16 (Rotated)
South Dealer
North-South Vulnerable

NORTH
♠ J 6 5
♥ K 10
♦ K 5
♣ K J 9 6 5 2

WEST
♠ K 10 8
♥ J 8 7 2
♦ 9 7 6 3
♣ 10 4

EAST
♠ 4 3 2
♥ 9 5 4
♦ A Q J 10
♣ A 8 7

SOUTH
♠ A Q 9 7
♥ A Q 6 3
♦ 8 4 2
♣ Q 3

West	North	East	South
			1♦
Pass	2♣	Pass	2NT
Pass	3NT///		

North-South had 25 high-card points and lots of clubs, a reasonable game try by most any standard. South (Katz) was declarer. West led off with his fourth heart and the dummy ♥10 held the trick, East discarding the ♥5. Katz naturally went after the club suit, and East played his ace on the third round. Time was running out for the defense. East tried to get to partner via a spade (not much chance to signal a suit preference holding the four, three and two); at any rate it was the only hope for the defense; and Katz had to finesse; otherwise he would not be able to unblock hearts. On lead again, West led . . . yes, another spade and Katz had his contract plus two. The West defender had three chances to lead a killer diamond but could not see beyond the shadows.

* * * *

WORLD TEAM OLYMPIAD - 1996. In France vs. Austria, both teams pushed a 24 high-card deal to 3NT. North's good intermediates, two jacks and three tens, apparently made the game try worthwhile.

Board O1-10
Dealer East
Both Vulnerable

NORTH
♠ K Q 10
♥ 10 9 6
♦ J 10 6
♣ K Q J 7

WEST
♠ 9 7 3
♥ A K 7 5 3 2
♦ 9 4
♣ 8 3

EAST
♠ A J 8 6 4 2
♥ - - -
♦ A 8 7 5
♣ 9 5 2

SOUTH
♠ 5
♥ Q J 8 4
♦ K Q 3 2
♣ A 10 6 4

France's Mari (North) declared in 3NT. Not wanting to concede a spade trick, East started with a low diamond. Now Mari could see four clubs, three diamonds and a spade; needing a slow heart or a second spade. However spades were at risk. East took the third diamond continuation and returned a club. Mari set out to find a heart trick, but when West won and returned a spade the game was up. Mari had to lose the ♦A, two hearts and two spades - down one.

In the other room, with North in the same 3NT contract, East led a diamond, but switched to a low spade after taking the second diamond with his ♦A, allowing declarer to make the contract. This was a 13 IMP swing to Austria.

Here we had notrump game tries with 24 high-card points (neither with a five-card suit), one making, the other not; the results depending on whether or not East aided with a spade lead, a choice made in the shadows.

*   *   *   *

BERMUDA BOWL - 2007. Despite Netherlands astute play of this board in the semifinals, Norway won the match and went on to defeat the USA team for the championship.

Board 55
Dealer South
All Vulnerable

NORTH
♠ Q J 7 6
♥ J 3
♦ J 9
♣ 9 6 5 4 2

WEST
♠ K 8 2
♥ K 8 7 6
♦ A Q 6 3
♣ 7 3

EAST
♠ 10 9 5
♥ Q 9 5 4
♦ 8 7 4
♣ Q 10 8

SOUTH
♠ A 4 3
♥ A 10 2
♦ K 10 5 2
♣ A K J

West	North	East	South
			1♣
Dbl	1♠	Pass	2NT
Pass	3NT ///		

You can see the problem South has trying to take nine tricks. It seems that the dummy has only a spade entry. This severely limits declarer's flexibility.

Netherland's Bakkeren was declarer. After taking the heart lead with his ♥A over East's ♥Q, Bakkeren played a low spade to establish his sole entry to dummy. West came up with the ♠K immediately and cashed his ♥K. Then, not being able to read anything useful from the heart discards, he elected to lead a small diamond. Declarer play the ♦J from dummy and - - - behold his second entry to dummy appeared out of the fog.

Bakkeren made the most of his good fortune. When the smoke cleared, he had his game plus two overtricks.

* * * *

We reviewed five deals where declarers needed help to make their notrump games. It is tempting to say the defenders erred; but that is too easy and does recognize the difficulty of finding the winning line of play. Nor does it credit declarers who disguise their problems and distract their opponents.

Defenders play in the shadows, where visibility fluctuates from board to board. It is the nature of the game that declarers are more equal, often winning tricks they have no way to obtain without help from their reluctant opponents.

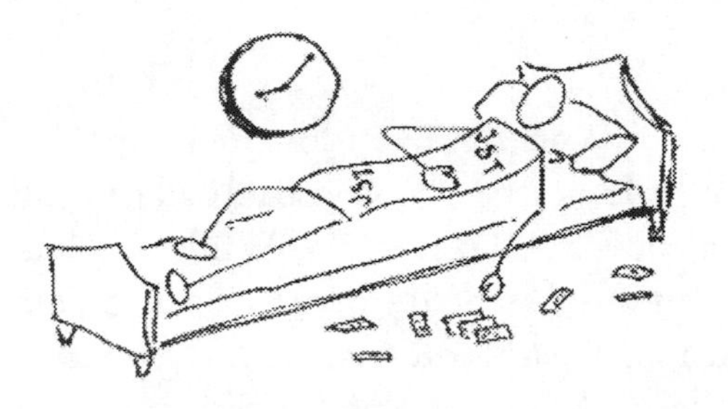

4

EARLY TO BED

So many bidding rules; they can and do overwhelm novices (maybe seniors too). Catchy phrases, like "drop dead" help. We all know what this means, even though the phrase is nearly obsolete. At least it is obsolete for those who employ Oswald Jacoby's transfer bids. Before transfers, any time you wanted to play in a suit at the two-level after partner opened 1NT, you simply bid that suit and partner knew better than to continue bidding (or else she would be history).

Here the focus is on situations where, after partner opens 1NT, you need to decide whether to play a part-score contract in your suit or in partner's notrump, assuming of course your system enables you to do this. These are deals where partner opens 1NT and you are too weak to consider game; but you have a suit that might play better than notump. Why is this an issue, you might ask? This issue is alive because pundits continue to differ in their advice; at least until now.

How many match points have you lost making 1NT exactly while others were making two hearts? You probably don't know; these deals don't happen often enough to keep track. While this is a ho-hum issue in rubber bridge, it is of substantial import in match-point where every deal is crucial; and a small part-score margin can produce as many match points as a grand slam. To win at match points you must score well in part-score deals because they account for more than half of all random deals. Simply put, you must get the most out of part-score deals or you will be a perpetual "also ran".

Partner opens 1NT and you have little or nothing to contribute except a five-card major. Looking only at your hand, can you predict the better choice between passing and transferring to your major? No, I venture to say you cannot because the answer is obvious only to those who have a preconceived notion.

*   *   *   *

Here is one of these part-score deals which highlights the choice between notrump and a part-score major.

Dealer: North
Vulnerable: None

NORTH
♠ A 5
♥ K 6 2
♦ A J 9 5
♣ A J 10 9

WEST
♠ Q 9 4 2
♥ A 10 5 4
♦ K 10 2
♣ 7 3

EAST
♠ K 7
♥ J 9 8 7
♦ 8 7 6 3
♣ K 8 6

SOUTH
♠ J 10 8 6 3
♥ Q 3
♦ Q 4
♣ Q 5 4 2

You are holding the North hand, a balanced 17 high-card points, and quite naturally open 1NT. South has one of those hands you love to hate, seven high-card points and five middling spades. South passes, leaving you to play 1NT.

The lead from East is the ♥7. West takes dummy's ♥Q and returns a heart to your ♥K. Now what? You see three clubs, two diamonds, one heart and a spade to make the contract. This may not be good enough because a 2♠ contract looks good. It may be tempting to try to establish some spade tricks, but the defense then will likely get two spades, three hearts and a club, holding you to seven tricks. Alternatively it appears that you might take six tricks in the minors and two in the majors if you could get to the dummy twice. Anyway, it is a close call as are so many of such deals, where the outcomes depend on the defenders skills, or yours.

*   *   *   *

Here's how some pundits have weighed in on this issue over the last several decades.

> In 1985 Charles Goren recommended passing partner's NT holding a near-pointless hand with a five-card major. This hand was given as an example:
>
> ♠XXXXX ♥XXXX ♦JXX ♣J
>
> In 1986 William Root advised to bid 2♠ (to play) following partner's 1NT opener holding a weak hand with a five-card or six-card major, such as:
>
> ♠975432 ♥3 ♦764 ♣862
> ♠QJ1097 ♥43 ♦Q74 ♣832
>
> However Root did not tell us when to pass partner's 1NT, if at all.
>
> More recently a respected pundit suggested that your choice, when responding to 1NT, should depend on the quality of your hand. Two examples were cited:
>
> 1. Go for the suit contract with the weakest of hands:
> ♠97654 ♥6 ♦10843 ♣942
>
> 2. Pass to play 1NT with a balanced intermediate hand:
> ♠Q9765 ♥QJ6 ♦J8 ♣Q92
>
> The rationale here seems to be that the first hand is worthless except for its spade length whereas the second hand is likely to contribute as many tricks in notrump as in spades. This seems reasonable.

No doubt these experts would argue their respective viewpoints persuasively, each bringing extensive bridge experience and creditability to the debate. But they conflict, and so we need a refreshing look.

At the table we see thirteen cards and we know partner is balanced and has about sixteen high-card points. This meager information is not enough. What we are trying to do is maximize the number of deals where we land in the better contract; the better contract being the one that produces the higher score or lesser loss. The rule should produce success at least 60% of the time. (In rubber bridge this is not much of an issue as making a contract is the key objective, and earning 10 or 20 points more or less is only marginally relevant.)

I analyzed a large number of random deals that fit the part-score profile, comparing outcomes either way, 1NT or two of a major. All of these deals involved a weak hand with a five-card major opposite a 1NT opener. This research produced a simple rule, (always the best kind). It is:

> **No matter what specific cards responder holds, the odds favor playing in the five-card major rather than 1NT.**

To be sure, some responder hand configurations make a difference in outcomes. Characteristics that overwhelmingly favor suit contracts are lack of high-card strength and shortness (void or singleton); conditions that favor the major suit some 90% of the time.

Notrump contracts do best with stronger and more balanced hands, but even then the odds are nearly 2:1 in favor of suit contracts. The advantage of a 5-2 fit or better is sufficient to produce on average at least one extra trick as compared to playing notrump.

There is no discernable hand characteristic that would sway the odds to favor notrump over the five-card major.

Generally the differences are modest (such as 110 points from eight tricks in a major compared to 90 points from seven tricks in notrump) but crucial. So when you have a five-card major and are too weak to invite game, bid two hearts or two spades to play, or transfer and pass, whichever is your style.

Using the transfer to direct partner into you major suit is the recommended practice because there is no risk of partner continuing to bid onward and upward. Moreover the transfer keeps the strong hand hidden from the defense, an event that occasionally earns its reputation.

*   *   *   *

Dealer: North
North-South Vulnerable

NORTH
♠ A 9 4
♥ K J 10 6
♦ A 10
♣ A 10 9 8

WEST
♠ Q 2
♥ A Q 2
♦ J 9 8
♣ K Q 7 6

EAST
♠ K 10 7
♥ 8 7 5 3
♦ Q 7 3 2
♣ J 4

SOUTH
♠ J 8 6 5 3
♥ 9 4
♦ K 6 5 4
♣ 5 3

North opens 1NT. South has minimal values and five spades, however anemic. If your partner holds this South hand, he can do you a favor by leading you into spades, initiating a transfer so that you become declarer in 2♠ with East on lead.

East leads the ♣J. This is not the easiest hand to play but it is no harder than notrump. You have only four quick tricks; hearts are at risk; and you are missing three key trump - king, queen and ten. Rather than draw trump early, the better line of play is to take the top two diamonds then ruff diamonds in your hand. You might bring in nine tricks but eight is enough because 1NT takes only six.

* * * *

In theory, the same notion (playing in a five-card suit rather than notrump) applies when responder has a five-card minor. Practically when playing Stayman and transfers, you cannot play 2♣ or 2♦, so pass to play 1NT. You might consider playing clubs or diamonds at the three level, but not without a six-card suit.

Here is a beggar's choice if there ever is one.

SOUTH
♠ Q 10 4
♥ 9 6 3 2
♦ Q 9 8 5 2
♣ 3

Next to nothing! Partner, though, is relatively wealthy and opens 1NT. East passes to you. We know that a hand such as this more often than not plays better in a five-card suit at the two-level than in 1 NT. In addition to five diamonds, your singleton club adds value in a suit contract. Assuming your partnership plays Stayman and transfers, you can't call 2♦ to play, but you can employ Stayman. You call 2♣; if partner returns 2♦ you have found a reasonable contract. Alternatively if partner returns 2♥, you have a 4-4 fit, also a reasonable contract. What you probably don't want to hear is a spade return where you would have a Moysian[(1)] fit; which may or may not play better than 1 NT.

———————

(1) Named after Alphonse Moyse Jr. who advocated opening four-card majors and raising with three-card support.

Here is the full deal for the hand shown above. It was played in a 16-table duplicate game.

Dealer: North
Vulnerable: None

NORTH
♠ K J 7
♥ Q J 5
♦ A K J 7 4
♣ J 5

WEST
♠ A 8 5 3
♥ K 10 4
♦ 10
♣ 10 7 6 4 3

EAST
♠ 9 6 2
♥ A 8 7
♦ 6 4
♣ A K Q 9 8

SOUTH
♠ Q 10 4
♥ 9 6 3 2
♦ Q 9 8 5 2
♣ 3

Several tables played in 2♦, which produces an easy eight tricks - two spades, five diamonds and a club ruff by South – and perhaps a heart trick as well. However, most pairs played in 1 NT, going down two tricks. The diamond players earned 10 match-points while notrump players earned only 5 match-points.

* * * *

Maybe the last deal is too unusual. You have to wonder how often you get a hit like this in diamonds. Here is the math. You get a diamond return 27% of the time, about 1 in 4. Otherwise partner returns hearts or spades showing a four-card (or five-card) suit. Naturally half of these are hearts and you then have a 4-4 major fit 36% of the time. You get to play in a preferred suit (diamonds or hearts) 63% of the time. The remaining instances you end up in 2♠ playing a 4-3 trump suit which still has the potential, given the singleton club, of producing one more trick than notrump. In those instances where both 1NT and 2♥ or 2♠ make, the major giving you 20 points more; and then any time the notrump contract is set, you come out ahead in the suit. Notably the singleton club is the most valuable asset.

Three-level contracts usually play quite well and score better than notrump when the weak responder has a six-card suit, so these are worthwhile if your partnership has a way to sign off at 3♣ or 3♦ (minor suit transfers). Frankly I would be happy to play 3♦ with six diamonds and no honors opposite partner's 1NT opener. One the other hand, if my suit is only five long, the chances of taking nine tricks with just half or less of the honors and maybe a 5-2 diamond fit are rather remote, whereas seven tricks in notrump, while also marginal, looks a lot better (or less worse as the case may be). Weak hands with a five-card suit are almost sure losers when you need nine tricks.

* * * *

The great advantage of standard notrump openers is that responder: (a) immediately knows the deal's potential within one trick; and (b) has the tools to determine the best strain and steer the bidding into it.

Now you know confidently where to rest when the deal needs to be put to bed early.

5

NUDGING THE ODDS

Card games, bridge in particular where the distribution of cards is random, we trust, lend themselves to the application of statistics. Random events within a limited universe like fifty-two cards can be predicted accurately, not by individual event of course, but over a large number of events - much like a random coin toss where you get either heads or tails, unpredictably toss-by-toss but precisely predictable at 50/50 over some multiple number of tosses.

Every solvent poker player has a working knowledge of card odds. Knowledge of these odds is even more crucial for the serious bridge player. Poker players have mere money at stake, whereas bridge players play for honor, glory, trophies and - - well whatever.

The odds of drawing a king from a full deck are four in fifty-two, one in thirteen. Now suppose you deal thirteen cards face up, and they contain two kings. The odds of drawing another king are two in thirty-nine, roughly one in twenty. That is simple math. The more cards you see, deduce and discover, the more you know about the rest of the cards you cannot see. Hence the better you can determine the wining line of play. Don't forget to ask yourself why the dog didn't bark!

The notion that the luck of the cards, good and bad, runs equally to both sides is true but irrelevant. Experts create their own luck, and it adds up to much more than a coin toss. A successful end-play changes odds from 50/50 for a finesse to a certain 100%. As Victor Mollo once wrote:

> **When I take a 50-50 chance, I expect it to come off eight or nine times out of ten.**

*   *   *   *

BERMUDA BOWL - 1999. It was the final match with USA1 playing Brazil. Three-quarters through the 160 board match USA1 had accumulated twice as many IMPs as Brazil, a margin of 443 to 224; yet they did not slacken the pace.

Board 137 (Rotated)
Dealer West
North-South Vulnerable

NORTH
♠ A J 10 8 5 2
♥ A 6 3 2
♦ J
♣ 8 7

WEST
♠ Q 9 6 4 3
♥ 8
♦ 6 5 3 2
♣ K 9 5

EAST
♠ - -
♥ J 10 9 7 5
♦ A Q 8 4
♣ A J 6 2

SOUTH
♠ K 7
♥ K Q 4
♦ K 10 9 7
♣ Q 10 4 3

West	North	East	South
Pass	1♠	Dbl	Redbl
Pass	2♠	Pass	3NT ///

Meckstroth was South. With game values opposite North's opener, he chose notrump rather than the 6-2 spade fit. East announced his strength and shape with a takeout double.

This would seem to be an easy contract if declarer could get five of the six spades in dummy. However as you see, West had all of the missing spades and Meckstroth had only one entry to dummy after the second spade was played. His prospects looked rather bleak.

West (Chagas) on lead, started with the ♣5 to partner's ♣A, and won the club return with his nine. Chagas switched to his lone heart, taken in Meckstroth's hand. Meckstroth began on spades; first the ♠K and finding the 5-0 split, continued to win the second spade with the ♠8. Three more spades tricks could be established by conceding the queen to Chagas, but Meckstroth could not afford it - the defense already had two clubs and potentially two more quick tricks, the ♣K and ♦A. So losing the ♠Q surely would set the contract.

Meckstroth led the singleton ♦J from dummy, taken by East's ♦A. The stage was set. East exited with a heart; Meckstroth took the ♥A then the ♥K; and threw Chagas into the lead with a low club to the ♣K. Chagas was end-played. He could give Meckstroth all of the spades or lead diamonds through his partner. Result: game and 600 points to team USA1.

In the other room Brazil's play of this same deal was anti-climatic. They played a 4♥ contract with a 4-3 fit and were set three tricks.

*   *   *   *

Surely you want to be in game with twenty-eight high-card points, but that considerable strength does not assure making it; the Fates who deal bridge hands are too capricious. There are several good possibilities to save you from the enigma of watching the defense run a string of tricks in your anemic suit. Maybe they will lead some other suit; it happens all the time. Or you can hold up playing your stopper until RHO has no more; a fundamental tactic for notrump declarers. But when all else fails, there is the squeeze.

* * * *

VENICE CUP - 1999. In the round robin stage we find Great Britain's McGowan making good use of a long diamond suit to make her notrump game in a match against Germany.

Board 14 (Rotated) Dealer North None Vulnerable	**NORTH** ♠ K Q 7 ♥ A K 8 7 ♦ Q 9 8 ♣ 8 6 3	
WEST ♠ 9 6 4 3 ♥ 10 5 2 ♦ J 2 ♣ 9 7 4 2		**EAST** ♠ 5 2 ♥ J 6 4 3 ♦ A 10 ♣ K Q J 10 5
	SOUTH ♠ A J 10 8 ♥ Q 9 ♦ K 7 6 5 4 3 ♣ A	

McGowan, South, was declarer in 3NT. West began by leading a low club; ta-ta stopper. McGowan had eight quick tricks and an easy diamond - that is after East wins the ♦A and runs four clubs. What do you do in this situation? Yes, run your solid suit. McGowan played out four spades. East followed Two times then discarded the ♦10. On the fourth spade, East gave up a good club to protect her heart jack: but to no avail. McGowan played a diamond to East's ace; that was it. The defense took only three clubs and the ♦A.

*　　　*　　　*　　　*

VENICE CUP - 1999. Canada's Francine Cimon played flawlessly this problematic notrump game.

Board 16 (Rotated)
Dealer North
North-South Vulnerable

NORTH
♠ Q J 5 4
♥ 10 8 6 5
♦ A K 7
♣ 4 2

WEST
♠ 9 6
♥ K J 9 7 3
♦ 10 4
♣ K Q J 10

EAST
♠ 10 8 7 2
♥ 4 2
♦ Q J 8 3
♣ 6 5 3

SOUTH
♠ A K 3
♥ A Q
♦ 9 6 5 2
♣ A 9 8 7

West	North	East	South
	Pass	Pass	1NT
2♥	3♥	Pass	3NT ///

Cimon, South, opened 1NT and, encouraged by partner in spite of West's heart overcall, went for the 3NT game.

West led clubs, Cimon holding off until the third round. She could see eight tricks. A finesse for the ♥Q was out of the question; somehow she had to induce West to lead hearts. After winning with the ♣A, she played two top diamonds, two top spades, and then exited with her last club to West, who now had nothing left but hearts, and thus was forced to play a heart into Cimon's A-Q. Luck? Not at all. Remember West's overcall?

As West followed to two diamonds and two spades, Cimon had the full count and the end play was routine.

*　　*　　*　　*

WORLD OPEN PAIRS - 1994. How would you feel defending, and being the butt of a double squeeze play? Be thankful you were not sitting West in this deal, one that was referred to by the commentator as "a defender's nightmare". It was played by Sweden in a match against Americans Berkowitz and Cohen .

Board 23
South Dealer
Both Vulnerable

NORTH
♠ J 10 9 8 5 4
♥ A J 6
♦ J 8
♣ A 2

WEST
♠ A Q 6
♥ 10 5 4 2
♦ 6 4
♣ J 10 9 8

EAST
♠ K 7 3 2
♥ Q 7
♦ K 7 5 2
♣ K 6 4

SOUTH
♠ - -
♥ K 9 8 3
♦ A Q 10 9 3
♣ Q 7 5 3

West	**North**	**East**	**South**
			1♦
Pass	1♠	Pass	2♣
Pass	2♥	Pass	2NT ///

South (Auby) was declarer, and West led the obvious ♣J, taken by partner's king. A club was returned to dummy's ace. Auby ran diamonds until East took the fourth diamond with his king, West discarding two hearts. Now Auby ran the good hearts, while West finally gave up his club stopper to hold on to the ♠A, to no avail. Auby made three overtricks.

*　　*　　*　　*

End plays, squeezes, double squeezes! All of these are tactics to avoid finesses and create tricks that you cannot get any other way. If none of these tactics are available, then assume a key card is located where you need it and play accordingly. There is, after all, more than skill involved. Luck is always lurking around to play its pervasive role. So a word of caution per Ron Clinger in ***Guide to Better Card Play:***

> **Do not make more assumptions than necessary. It is easier to have one prayer answered than two.**

If you hope to bid and make lite notrump games, you must squeeze every possible trick out of each deal, and then perhaps one more. You can not just play according to the statistical odds, sometimes you have to nudge them in your direction.

6

MINORS

Minors will never come of age, no matter if the game is match-point, IMPs, or rubber. Two extra tricks to make game is simply too much of a deficit to overcome. It is quite uncommon to find a deal with enough honors to enable eleven tricks in a minor yet be unable to take nine in notrump.

There are of course certain kinds of hands that are unsuitable for notrump. A highly skewed distribution or a clear and certain Achilles' heel somewhere are, or seem to be, good and sufficient reasons to abandon notrump in favor of a minor. Sometimes, though, even those reasons are insufficient. When apparently unsuitable deals do appear, certainly not very often, even then sometimes notrump produces better results.

Look at the following deal where North-South hold 24 high-card points and an attractive diamond fit.

NORTH
♠ 8 7 4 2
♥ A Q 10
♦ Q 10 5 3
♣ 8 2

SOUTH
♠ K Q 5
♥ J 3
♦ A K 9 7 6
♣ Q J 4

South opens 1NT and North raises to 2NT. Assuming you are playing the South hand, you may stop at 2NT or continue to game. In view of your attractive diamonds, perhaps you should continue.

Already you can count five diamonds, two hearts, and one spade for eight tricks. The ninth could come from another spade if West has the ace, or a third heart if West has the king. A club lead from West, setting up your queen of clubs, might do it. There are so many possibilities. What you get is a low spade to your queen.

You need to finesse hearts, but if you run diamonds first, you will no doubt cause both defenders to make tough discard decisions. Anyway it is a good game try, 50/50 plus a chance of defense error.

* * * *

NORTH
♠ K J 4
♥ 9
♦ K 9 6 5 4
♣ A K 10 3

SOUTH
♠ A 7 3
♥ Q J 8 5 2
♦ Q J 10 7
♣ 9

West	North	East	South
			Pass
Pass	1♦	Pass	1♥
Pass	2♣	Pass	??

Neither hand is balanced, but notably they have a 5-4 diamond suit. You, South, are dealer.

Partner's bidding shows 5-4 in the minors. You have diamond support and a good 10 high-card points. Do you raise diamonds? There is an outside chance of a 5♦ game. On the other hand, you have stoppers in every suit so why try for 11 tricks when the same luck (maybe even a smaller dose) might produce nine tricks in notrump. If you bid 2NT, will partner pass? Let's assume that North continues to 3NT.

West leads the ♣4. You have two sure clubs, four diamonds after the ace is extracted, and two top spades; eight tricks and a good prospect for another spade, another club or even a slow heart. If West has four clubs including one honor, you can be certain of a third club trick. Surely this is a worthy notrump game try.

*　　*　　*　　*

BERMUDA BOWL - 1991. Poland was playing Iceland for the championship. Here is a deal where East-West had a very good 6-2 diamond fit and 25 high-card points. Are these hands good enough for game?

Board 38
East Dealer
East-West Vulnerable

WEST	**EAST**
♠ A 7 6	♠ K
♥ 9 6 3	♥ A 10 2
♦ J 3	♦ A K 10 7 5 2
♣ K Q 7 6 2	♣ J 9 8

West	**North**	**East**	**South**
Poland	Iceland	Poland	Iceland
		1♦	2♦*
Dbl	4♠	5♦ ///	

* hearts & spades

Here in the closed room, South interfered with a conventional call and North followed with a pre-emptive 4♠. This made it impossible for Poland to play in 3NT if they had been so inclined. At any rate, Szymanowski (Poland) made his 5♣ game when he found the ♦Q onside.

Meanwhile in the open room, Iceland coasted into 3NT as East bid diamonds twice and North and South remained silent.

West	**North**	**East**	**South**
Iceland	Poland	Iceland	Poland
		1♦	Pass
2♣	Pass	2♦	Pass
2NT	Pass	3NT ///	

Perhaps some would have passed 2NT; but the prospect of diamonds bringing in five or six tricks easily justifies the game try. Anyway, Iceland's Jonsson won eleven tricks for plus 60 points over the diamond game.

*   *   *   *

BERMUDA BOWL - 1993. You may feel insecure playing in notrump while having just one stopper in a suit that had been bid and supported by the opposition. But it so happens that most of these turn out well if you have full values elsewhere.

In the round robins, sixteen teams competed for eight places in the quarterfinals. Norway and Denmark are featured here.

Board 19
South Dealer
East-West Vulnerable

WEST	**EAST**
♠ J 4 3	♠ Q 2
♥ 10	♥ K 9 5
♦ K Q 10 8 5 2	♦ A J 9 7 4
♣ Q 8 2	♣ A J 6

West	**North**	**East**	**South**
Norway	Denmark	Norway	Denmark
			Pass
Pass	1♥	1NT	2♥
3NT	Pass	Pass	Dbl ///

East (Norway's Aa) was not intimidated by North's opening heart call; with 15 high-card points he overcalled 1NT. Partner, with eight points and a super diamond suit, went for the notrump game, and was promptly doubled by an irked South.

South naturally led a heart to Aa's king; so Aa needed only the club finesse for his ninth trick; a 50/50 chance that favored declarer (because North had most of the outstanding honors). This game earned Norway 750 points and 12 IMPs.

Even with eleven, diamonds never got into the auction, and just as well since five diamonds had no chance.

*   *   *   *

In game contracts of notrump vs. a minor, notrump wins hands down. Needing two more tricks for game, the minors cannot compete except where there is a fatal weakness in notrump.

Scoring rules for slams alter the balance. Here minors require the same number of tricks as notrump, and the slam bonus is the same for both strains. There remains the difference in trick values which favors notrump by 70 points, a full-size margin at match points but a ho-hum difference in rubber or IMPs.

It should be self-evident that wherever there is an equal prospect for making slam, the nod should go to notump. Beyond that the focus should be entirely on viability because the slam bonus is so large that no one would choose the higher-risk contract over the lesser one.

*   *   *   *

VENICE CUP - 1991. USAII and Austria reached the final of 128 boards. Well into the match where the Americans held a large lead, Deas & Cohen (USAII) held these strong hands.

Board 105
North Dealer
East-West Vulnerable

NORTH
♠ A 10 6 3
♥ A K 5
♦ A J 8
♣ K 9 4

SOUTH
♠ K Q
♥ 8 6 2
♦ K Q 10 7 5 4
♣ 3 2

West	**North**	**East**	**South**
	Deas		Cohen
			Pass
Pass	1♣	1♥	2♦
Pass	2♥	Pass	3♦
Pass	4♦	Pass	4♥
Pass	6NT ///		

Deas and Cohen drove to the notrump slam based in large measure on the solid diamond suit. East led a heart to Deas' ace. She had eleven tricks in the bag, needing one more, which had to come from the fourth spade or ♣K. Deas tried three spades, but the ♠J did not fall. Then It was too late to try for the ♣K. Down one.

The odds favor playing for the ♣K, (barely), but you have to maintain control of the other suits in case West comes up with the ace of clubs immediately. If Deas had tried clubs early, she could have been set several tricks; whereas trying for the fourth spade risked only a one-trick set. In the other room, the Austrians stopped in 3NT - an easy contract.

This deal was also played in a minor slam (6♦) by both teams in the Bermuda Bowl. The result with regard to minor vs. notrump slams was inconclusive as one team made 6♦ while the other was set one trick.

* * * *

The advantage that minors cling to at slam level is not self-evident. Logic suggests that highly unbalanced deals are safer in suit contracts than in notrump. But logic is not enough. The difficulty here is that we don't see or play enough relevant deals; and those we do encounter usually are not subjected to rigorous analysis regarding the potential results from the untried strain. In other words, having made a 6♣ contract for instance, we do not routinely, or hardly at all, re-examine the deal as to the probable consequence of contracting for 6NT.

This is a vexing question which needs a more definitive answer. To that end, I sifted through hundreds of thousands of random deals to find and study those few that were relevant to this issue. The process was tedious and rather dull, so let's cut through the chaff to the conclusions.

*　　*　　*　　*

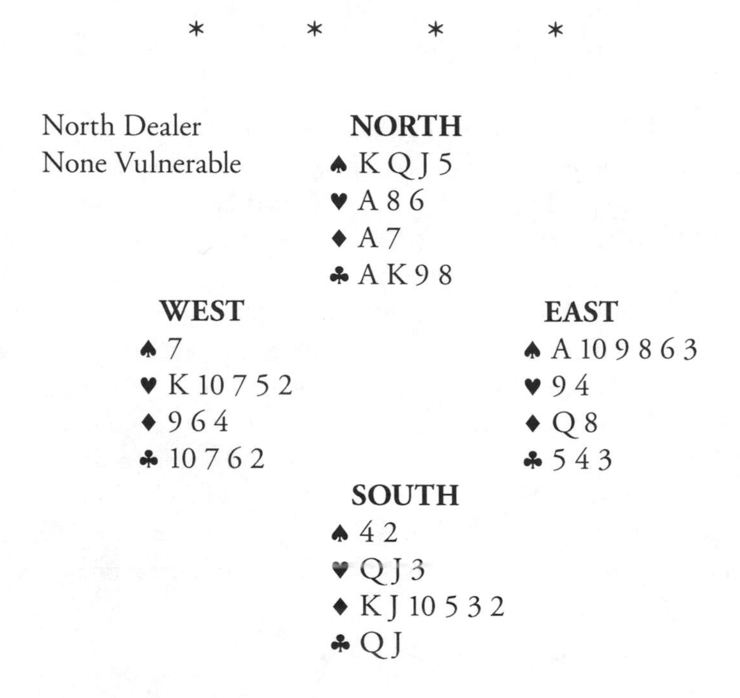

In a diamond slam, West leads the ♠7, and - - - jackpot; a spade to the ace, and a spade return for West to ruff, and the contract is off one before it hardly began.

When North plays 6NT, there are 12 tricks to be had no matter what East leads: declarer concedes the ♠A and claims. The score is +1040, 14 IMPs, a top board, in favor of notrump.

*　　*　　*　　*

NORTH

North Dealer
East-West Vulnerable

♠ Q 7 5
♥ K 5
♦ A K Q 10 5
♣ A J 4

WEST
♠ K 10 9 8 6 2
♥ Q 8 6
♦ 9
♣ 9 6 3

EAST
♠ J 4
♥ J 9 7 3 2
♦ 6
♣ Q 10 8 7 2

SOUTH
♠ A 3
♥ A 10 4
♦ J 8 7 4 3 2
♣ K 5

North playing 6NT has eleven quick tricks. If East leads his fourth club, it gives declarer the ♣J and slam is a dunk. Instead, let's assume East leads the fourth heart following the theory that if there is an unbid major and an unbid minor the major is the better lead, however marginal. The odds of making this contract are quite good. North plays from dummy (South) to the ♠Q early. If this fails, he will have to finesse clubs. Thus we see that if West holds either the ♠K or the ♣Q, declarer makes the contract. Two 50/50 chances equal 75/25.

Assume North, having opened 1♦, is declarer at 6♦. East perceptively leads a heart, which does no damage. Declarer draws trump and later ruffs a heart for the 12th trick. Along the way, he can try the club finesse for a 13th trick without risk.

Now for the math (sorry, it's a bit complicated). The diamond contract brings in 930, including 50% credit for the 13th trick. The notrump contract at 75% odds brings in on average 730 points. The statistical margin in favor of the minor is 200 points or 5 IMPs.

* * * *

The two deals presented above are outliers of what you can expect in minor vs. notrump slams. The immediate question is: which is the better contract on average, 6NT or 6 of a minor? By better I mean the one producing the most points, not the one you would rather play!

In slam contracts, over 90% of the relevant deals are ties in the sense that either contract takes the same number of tricks. Since scoring rules favor notrump to the extent of 10 or 20 points per trick, notrump should be the preferred contract, at least in match-point duplicate.

Deals where the potential tricks differ sometimes favor notrump and other times favor minors. These deals are scarce and unpredictable; unpredictable in that they are beyond the ability of most of us to find the better contract.

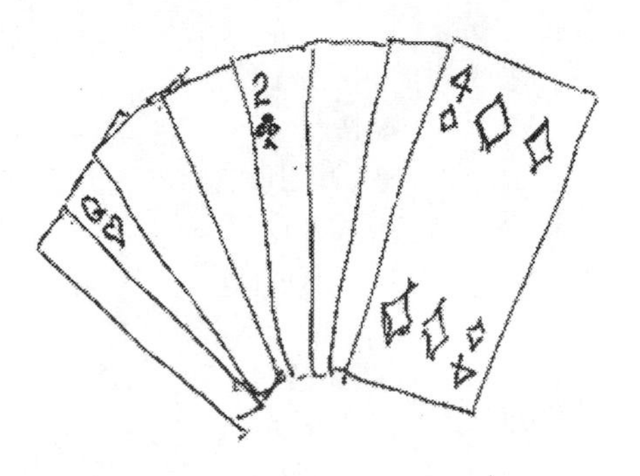

7

EMPTY SPACES

The term "empty spaces" is a euphemism for unknown or unidentifiable cards in the opponents' hands. For instance, when West opens one heart you can count on him holding at least five hearts. That tidbit also produces other useful information: it limits West's holdings in other suits to eight; then when you add your hearts and dummy hearts to West's five, you have a good fix on East's hearts. All of that from one bid; and you can do a similar placement of high-card values from the same bid. Then when left-hand opponent selects an opening lead . . . .

The objective of course is to figure out what you need to know of suit and honor distributions between the two opponents' hands in order to maximize your score.

You are South, and the North-South hands are:

NORTH
♠ A 4
♥ Q J 10
♦ K 9 6 5
♣ A K Q 3

SOUTH
♠ K 7 2
♥ 8 4 3
♦ A J 10 3
♣ 6 5 4

West	**North**	**East**	**South**
Pass	1♦	Pass	1NT
Pass	3NT ///		

West leads the ♠5 which you duck, allowing East to win with the queen; then you take the spade return with the dummy ace. You have seven tricks off the top. You cannot try for a slow heart as the defenders would win two hearts and three spades, assuming, as is quite likely, West began with five spades. You need to find clubs divided 3-3 or win all four diamonds, finessing for the missing ♦Q. Fortunately you can finesse diamonds through either defender.

The opening lead indicates West began with five spades thus eight empty spaces (unknown as to suit), while East has three spades and ten suit unknowns. Hence the odds are East has more diamonds than West by a small margin. This is hardly enough information yet to be useful.

You discover more by playing clubs. Suppose everyone follows two club tricks. Now you play a third club. If both follow, you win the 13th club and the diamond finesse might give you an overtrick. Alternatively, If you find that East had just two clubs, you have narrowed down the empty spaces: You now know than East has seven red-suit cards and West has four. The probability of East having the missing ♦Q is 7/11, nearly 2:1, and the best play is to finesse diamonds through East.

*　　*　　*　　*

WORLD TEAM OLYMPIAD - 1992. USA vs. France, half way through the 96 board championship match:

Board 51 (Rotated)
Dealer East
North-South Vulnerable

	NORTH	
	♠ Q 9 7 6	
	♥ 8 4	
	♦ K 10 7	
	♣ K 10 7 2	
WEST		**EAST**
♠ A 10 3 2		♠ 8 5
♥ K Q J 6 5		♥ 7 2
♦ 9 8 5 2		♦ Q 4 3
♣ - -		♣ A J 9 8 5 4
	SOUTH	
	♠ K J 4	
	♥ A 10 9 3	
	♦ A J 6	
	♣ Q 6 3	

West	North	East	South
		Pass	1NT
2♣*	2NT	Pass	3NT ///

* Hearts & spades

Meckstroth (South) was declarer at 3NT, West having advertised his majors along the way. West led his ♥K then his ♥Q; Meckstroth took the second trick and was assured that East had begun with two hearts. He played the ♠K to West's ♠A. Running out of options, West played his remaining high heart and exited with a spade, dummy ♠9 winning. (There was no point to a heart play to establish his fifth, because he had no entry left.) Meckstroth came to his hand with a spade and played clubs to his king and East's ace. That was it. Either minor suit return would give Meckstroth his ninth trick.

West's conventional bid indicated he had eight or nine in the majors. Ergo, Meckstroth could figure him for four in the minors. This meant East had some nine minors. Early play of the majors confirmed this so Meckstroth had the sure end-play.

Elsewhere the French played in 3♠, making, after Hamman opened a preemptive 3♣. Bottom line was a gain of 10 IMPs for USA.

*   *   *   *

WORLD TEAM OLYMPIAD - 1996. Experts can be excused for missing the best line of play once in a while, especially when it involves intricate play sequences or complex card combinations. That it doesn't happen very often is a great tribute to their expertise. This deal was played in the final between France and Indonesia.

Board 29 (Rotated)
Dealer East
Both Vulnerable

NORTH
♠ K 4
♥ K Q J 8 4
♦ A 10 3
♣ K 5 2

WEST
♠ J 7 2
♥ A 10 9 7
♦ 4
♣ J 9 8 7 6

EAST
♠ 9 8 6 5 3
♥ 6 5
♦ K J 7 6
♣ A Q

SOUTH
♠ A Q 10
♥ 3 2
♦ Q 9 8 5 2
♣ 10 4 3

South was declarer in 3NT, East-West passing throughout. West led the ♣7 to the ♣2, ♣Q and declarer's ♣10. Now East had the obvious problem of not being able to continue clubs without setting up declarer's king. So he shifted to a low spade which declarer took with the dummy ♠K.

South could be sure of three spades and two hearts, thus needing four diamonds. South began with the ♦A from dummy then low to the ♦Q. This was a reasoned sequence, and it would win (losing only one diamond) whenever the queen won and diamonds were divided 3-2, or when an honor fell under the ace or queen. This is a 60% play, quite good for a game try; but alas in this particular instance declarer could not bring in the diamonds and at the end he was off three tricks.

West had led what looked like his fourth club, and partner had declined to return the suit. The most likely explanation was that East had either a singleton or the ♣A and ♣Q. West had at least five clubs and East had more empty spaces than West. This information suggests playing for a diamond split of 3-2 or 4-1 with East holding the majority. In such a scenario, your best odds are to play East for the ♦J, finessing twice if necessary. When East has more diamonds than West, this sequence raises the odds in declarer's favor by 10 percentage points.

* * * *

Determining the characteristics (suits, honors, spots) of empty spaces is a vital way to beat the statistical odds. You don't have to know the math or calculate odds at the table. What you need is to recognize that missing key cards are more likely to be in the hand with the most cards in that suit. Say there are five spades outstanding and clubs are known to be divided 5-2 with five in the West hand. Absent any other information, you would play East to hold more spades than West; and of course East is more likely to hold any particular spade honor as well.

There are three common and essential sources of data that enable you to fill in empty spaces in the opponents' hands.

> **Bidding:** Every call by an opponent reveals something about the hand, including "pass". A "pass" denies sufficient honors to bid; and most likely also means the hand does not have a seven-card suit or even, in these days of light preempts, a six-card suit.

Opening Lead: It is amazing how much an opening lead can tell you. Don't forget to ask yourself not only why that particular suit and card was led, but also why another suit was not selected instead.

Discovery: It should be your routine practice to delay making a key play decision as long as possible, not, I hasten to add, by slow play, but by first playing tricks that are not critical to the outcome.

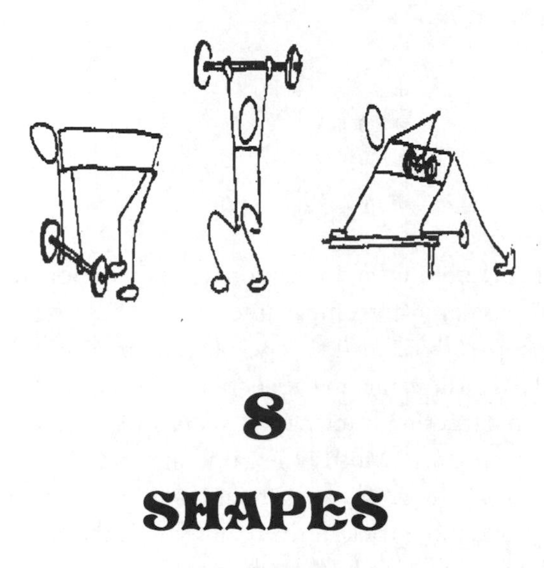

8
SHAPES

The notions of balance and notrump have always been linked. From the beginning (which for a few of us was Goren and fewer Culbertson) notrump openers were by definition balanced, and balance was 4-3-3-3, 4-4-3-2, and somewhat reluctantly 5-3-3-2. Nevertheless the distributional diversity of deals that end up, or should, as notrump contracts will surprise all but the most progressive tournament bridge players.

Other than hands that are clearly meant for notrump - that is balanced with abundant trick-takers and generally lacking an attractive major suit - notrump becomes the repository of last resort for deals that lack a playable trump suit. This is not necessarily a bad consequence; mostly these notrump contracts are good even though they may be bid as a last resort. It would not be unreasonable to view most of such deals as favoring notrump, a view quite different than as a fall-back or last resort choice. No matter how you choose to view them, they produce surprises, mostly pleasing for the declarer.

The most widely recognized notrump game try with a skewed hand is the "gambling notrump":

♠ 4
♥ Q J 3
♦ Q 5
♣ A K Q 10 6 3 2

This hand does not fit the conventional notion of notrump, nonetheless it is the basis for many notrump game tries. These hands have a long, running suit (hopefully) and little else but a stopper in whatever suit the opponents had bid. The earlier in the auction it is called, the better to prevent the defense from garnering information about strengths and weaknesses of your side or of their own. Mostly the gambling call is employed where the strong/long suit is a minor. With a similar holding in a major, generally it is better to try for one more trick in the major suit as the risk, not to mention the embarrassment, of losing badly is greatly lessened.

*   *   *   *

BERMUDA BOWL - 1995. Here in the championship match between Canada and USA2, North drew a club suit to die for.

Board 134
Dealer East
East-West Vulnerable

NORTH
♠ 4
♥ Q J 3
♦ Q 5
♣ A K Q 10 6 3 2

WEST
♠ A Q 6
♥ K 10 9 5
♦ 9 8 7 6 2
♣ J

EAST
♠ J 9 8 7
♥ A 8 7 2
♦ K 4 3
♣ 8 7

SOUTH
♠ K 10 5 3 2
♥ 6 4
♦ A J 10
♣ 9 5 4

West	**North**	**East**	**South**
		Pass	Pass
1♥	3NT ///		

Meckstroth (West) opened a heart and Silver (North) overcalled 3NT. On lead, Rodwell (East) was sure that Silver would have a heart stopper; so between spades and diamonds he chose a small diamond, hoping Silver would not expect him to lead away from a king. From Silver's view, the contract was not secure; he could lose a diamond, two hearts and two or three spades. At any rate, he did what had to be done, played the ♦10 to establish his ninth trick and hoped the defense would not find five quick tricks. Behold, Rodwell had lead from the ♦K and Silver's game was in the bag. In retrospect, Silver had no fatal weakness and his game was secure anyway.

In the other room, after three passes Wolff (North) opened 3NT. Hamman (South) decided to play safe with a take out to 4♣. (If Wolff's suit was diamonds rather than clubs, he would of course correct to that suit.) Four clubs was a good contract, but no match for the notrump game. Canada won seven IMPs.

*   *   *   *

BERMUDA BOWL - 1993. Sixteen teams survived a world zonal competition to compete for the championship. Here we have Chile and Venezuela competing in the quarterfinals.

Board 14 (Rotated)
Dealer South
Both Vulnerable

NORTH
♠ J 2
♥ K Q 7 6 2
♦ 7 6 4 3
♣ 6 3

WEST
♠ A 9 8 4 3
♥ A 10
♦ Q 10 9 8 5
♣ 10

EAST
♠ 10
♥ J 9 8 5 4
♦ K J 2
♣ Q 7 5 4

SOUTH
♠ K Q 7 6 5
♥ 3
♦ A
♣ A K J 9 8 2

West	**North**	**East**	**South**
			1♣
1♠	Dbl	Pass	3♣
Pass	3♥	Pass	3NT ///

In the other room the North-South (Venezuela) made a ho-hum 3♣. But here Chile South (Plaut) tried a notrump game lite in honors but with an impressive string of clubs.

West led a low spade, won by the dummy ♠J. Declarer started clubs; a low club to the jack and continued the suit until East finally won with the ♣Q, West discarding two diamonds along the way. The defense drove out the one diamond stopper and South could then win two clubs, but that was it. The defense won two diamonds, one club and the two other aces, for a one-trick set. Ah, for once lady luck was in the defense's court, having dealt them 4-1 in South's hoped-for running suit.

*   *   *   *

SENIORS BOWL - 2007. The first Seniors Bowl was played in 2001, the first three won by US teams. In the round robin, behind more than 100 IMPs, Italy's DeFalco went for 3NT. He had just nine high-card points opposite partner's twelve!

Board 68
Dealer West
Both Vulnerable

NORTH
♠ Q 10 6
♥ A J 10
♦ J 10 9 7 6
♣ J 3

WEST
♠ A K 9 4
♥ K 8 7 5 3
♦ 8 5 4
♣ 9

EAST
♠ 7 5
♥ 6 4
♦ A K Q 3 2
♣ 10 8 7 5

SOUTH
♠ J 8 3 2
♥ Q 9 2
♦ - -
♣ A K Q 6 4 2

West	**North**	**East**	**South**
Pass	Pass	Pass	2♣
Pass	3NT ///		

Opposite this unbalanced club opener, DeFalco (North), with some balance but just nine high-card points, desperately went for the game. The defense had available five quick tricks (three diamonds and two spades) if they could find them soon enough.

East led off with the ♦K followed by the ♦Q, partner following with the ♦4 and ♦8, suggesting a switch: hearts or spades? DeFalco had discarded two small spades from dummy, holding onto his hearts. Was this necessity, desperation, or guile? Whatever it was, it worked as East switched to a heart, and DeFalco had his quick nine tricks.

*   *   *   *

There are deals that can't be made because the cards are wrong or a suit is weak, but there is no such thing as a misfit. After many years of trying to find truths about notrump, I concluded that the notion "if you can't find a suit, bid notrump" is backward. It should be:

Run to a suit only when there is a fatal weakness in notrump.

Well then you might ask: "How do you know there is a fatal weakness?" Mostly you don't. The main difficulty is that very often a critical weakness as seen double dummy remains undiscovered by the defense, hence is not fatal.

There are some sign-posts to watch for, however. If your side has bid three suits, clearly you are vulnerable to attack in the fourth. Another is a bid by the opposition that is indicative of a five-card suit; be wary if you don't have a good stopper and some length there. Surprisingly a good sign is a misfit; if your side does not have an eight-card fit, it is nearly certain you have one or more stoppers in every suit (the underlying assumption being that your side has the requisite strength for the proposed contract).

There is another situation in declarer's favor. With a bit of luck the defenders will have exhausted their best suit after taking just four tricks.

*   *   *   *

VENICE CUP - 1993. It was the quarterfinals and two USA teams made it this far. Here USA1 was matched against Germany. North held spades, South hearts, and both diamonds, with copious honors between them.

Board 39
Dealer South
Both Vulnerable

NORTH
♠ A K 6 5 3 2
♥ - -
♦ K Q 8 5 4
♣ A J

WEST
♠ J 10 9
♥ K 10 7 5 3 2
♦ 7
♣ 8 6 5

EAST
♠ 8 7 4
♥ 9 4
♦ A 10 9 2
♣ 10 4 3 2

SOUTH
♠ Q
♥ A Q J 8 6
♦ J 6 3
♣ K Q 9 7

West	**North**	**East**	**South**
(USA1)	(Ger)	(USA1)	(Ger)
			1♥
Pass	1♠	Pass	2♣
Pass	2♦	Pass	2♥
Pass	3♦	Pass	3NT ///

In the closed room, the Germans settled into 3NT after a lengthy exploration where every suit was bid along the way. There were more than enough honors and multiple stoppers everywhere. In 3NT, West led a spade to South's ♠Q. South then lead a low diamond to the ♦K and ♦A, and with spades running, twelve tricks were in hand.

In the open room USA1 North (Deas) took partner out of 3NT presumably into a more secure 4♠. Deas won the heart lead in dummy and played the ♠Q to unblock the trump suit. Deas then played a diamond to drive out the ace. East immediately returned another diamond, giving partner a ruff and holding declarer to eleven tricks - one IMP to Germany.

Was this a biddable slam? There are 32 high-card points here, divided 15 and 17. These are intermediate hands which are often difficult to define strength accurately. The only fit is diamonds where a slam would be doomed to fail because of the 4-1 split. Alternatively slams in spades or notrump are feasible, but only because South's single spade is the queen.

*　*　*　*

No doubt you had been duly warned long ago to beware of the worthless doubleton when you are contemplating a notrump contract. If you think worthless doubletons are frightening, what about two of them in the same hand - that is hands of 5-4-2-2 distribution? Here we have several such deals. Not all two-card suits are worthless, but still they can be nail biters.

WORLD TEAM OLYMPIAD - 2004. Italy and Netherlands survived to the finals. Nearing the end Italy was well ahead, which might account for aggressive Netherlands bidding.

Board 137
North Dealer
East-West Vulnerable

NORTH
♠ K 10 7 6
♥ Q 7 4 3
♦ Q 6 5
♣ K 10

WEST
♠ 9 4 2
♥ 10 9 8 6
♦ A K 7 4
♣ 9 6

EAST
♠ A 8 5 3
♥ K J 5
♦ 9 2
♣ J 8 5 2

SOUTH
♠ Q J
♥ A 2
♦ J 10 8 3
♣ A Q 7 4 3

West	**North**	**East**	**South**
	Pass	Pass	1NT
Pass	3NT ///		

South (Brink, Netherlands) opened 1NT with this marginal 5-4-2-2 hand, and North bypassed Stayman in favor of notrump. West led the ♦A then switched to the ♥10, ducked all around. Another heart doomed the contract as the defense had two hearts, two diamonds, and the spade ace for a one-trick set.

In the other room, South (Nunes, Italy) also opened 1NT. After trying for a major fit, the pair ended in 3NT, also off one trick. (This Italian pair opens 1NT with all in-range 5-4-2-2 hands except when the 5-4 suits are majors.)

If you were to open the South hand 1♣ instead of 1NT, it is likely you would end in a notrump contract anyway.

*   *   *   *

BERMUDA BOWL - 2007. Norway won its first ever Bermuda Bowl, beating USA1 in the final contest.

Board 77
North Dealer
All Vulnerable

	NORTH	
	♠ J 8	
	♥ 7 6 5 4 2	
	♦ 8 6 5	
	♣ J 8 3	
WEST		**EAST**
♠ K 7 2		♠ 10 3
♥ 9		♥ A K Q 10
♦ Q 9 2		♦ A K J 7 4
♣ A Q 10 7 5 4		♣ 6 2
	SOUTH	
	♠ A Q 9 6 5 4	
	♥ J 8 3	
	♦ 10 3	
	♣ K 9	

In both rooms, East opened 1♦ and South overcalled 1♠. Thereafter the bidding took widely different paths and the Norway team ended in 5♦ while USA1 settled at 5♣. Both contracts made for a tie 600 points. Note however that 3NT is an easy contract with overtricks to boot.

*   *   *   *

SENIORS BOWL - 2007. Here we find a North American final with USA1 playing Canada. Both teams arrived at 3NT, but by quite different paths.

RR16, Board 2 (Rotated)
North Dealer
East-West Vulnerable

	NORTH	
	♠ K 10 8 ♥ 10 9 8 5 ♦ K 4 2 ♣ Q 8 2	
WEST ♠ Q 9 7 ♥ A Q J 6 ♦ 3 ♣ K 7 5 4 3		**EAST** ♠ J 4 3 ♥ 7 4 2 ♦ Q 10 9 8 ♣ 10 9 6
	SOUTH ♠ A 6 5 2 ♥ K 3 ♦ A J 7 6 5 ♣ A J	

West	North	East	South
	Pass	Pass	1NT
Pass	2♣	Pass	2♠
Pass	2NT	Pass	3NT ///

USA1 (South) opened 1NT with this 5-4-2-2 hand, and, after trying for a heart fit, pushed on to the notrump game with South (Kasle) declarer. A club was led to Kasle's ♣J, and he set out to establish diamonds: ♦K in dummy, then finessing the ♦Q and getting the bad news that East had a diamond winner. Kasle played his ♦A, and conceded the fourth diamond to East. A heart through Kasle and another club to dummy did the trick. Kasle could find only eight tricks, down one.

When the Canadians held the North-South hands, South opened 1♦ and the pair reached 3NT from the other side of the table with North (Baran) declaring. After a club lead to the ♣K and a return to declarer's ♣A, Baran was in essentially the same situation as Kasle had been in the other room. However Baran conceded the third diamond to East to avoid a discard from his hand, while West discarded the ♣3 then ♣4 on the second and third diamond tricks. If there was a suit preference in these club discards, East missed it as he led a spade, taken by West as Baran ducked. From this point Baran had three spade tricks and thus could not be beaten: making game and 10 IMPs to Canada.

The essential difference between making the contract or not was East's lead upon taking his sole diamond trick. A heart lead set the contract whereas a spade lead enabled making the contract. This was a very peculiar outcome considering that, when South was dummy with the ♥K ♥3 tabled, East chose to lead a spade; but when South was declarer holding the ♥K and ♥3 in his closed hand, East found the killing heart lead.

* * * *

	NORTH	
East Dealer	♠ J 10 5 4	
North-South Vulnerable	♥ A K 10 9	
	♦ 5	
	♣ 8 6 3 2	
WEST		**EAST**
♠ Q 9 8 3		♠ A 7 6
♥ Q 4 3		♥ 8 7 6 3
♦ Q 10 8 7 4 3		♦ 6
♣ - -		♣ A J 10 9 4
	SOUTH	
	♠ K 2	
	♥ J 5	
	♦ A K J 9 2	
	♣ K Q 7 5	

If South opens 1NT (2-2-5-4), the contract will be played in notrump. West leads the fourth diamond and there is no way to defeat the contract. Alternatively, if South opens 1♦, North says 1♥. Would you rebid two clubs, or 2NT? The point is that it can be more difficult finding a good notrump contract after a suit opener.

* * * *

Each of the preceding deals involved hands with 5-4-2-2 distributions. And in each notrump was a viable, if not superior contract. In two deals where notrump was not the opening choice, notrump openers would have produced better results.

In two instances, notrump was the opening choice with 5-4-2-2 hands. At first blush it seems that opening 1NT risks missing a good suit contract; and it would seem to double the probability of opponents running one of declarers' short suits if the bidding ends in notrump.

I turned to random deals to get a better notion of the viability of opening these hands in notrump rather than a five-card minor. Minors have that huge score handicap, so you might say I loaded the dice. I avoided five-card majors because my intuition tells me not to open notrump with an unbalanced hand containing a five-card major. Besides I would have to find a new partner.

In a nut-shell, the results were 14% in favor of minor openers, 29% in favor of notrump openers, and the remainder ties. There are three substantial reasons notrump wins out:

Most of the time notrump is the better contract; and when it is, notrump is the better opener.

Opening 1NT puts the contract in the strong hand with the opening lead coming to it.

Opening 1NT inhibits lead-directing overcalls.

Lead-directing overcalls are going to happen much more often when you open a minor than when you open 1NT. When they do occur, potentially they can be so lethal that you are compelled to retreat to the minor.

9
HONORS UP

Honors are where you find them, if you can. The import of empty spaces in sorting out defenders' holdings has been noted. Other key bits of information to seek out are locations of key honors. How are honors divided between the two hands? Who has the crucial king, or ace, or whatever? Most of the time you cannot be sure of such information so you have to view it in terms of probabilities: which opponent is likely to have the key card you are concerned about. When you combine clues regarding missing honors with the implications derived from empty spaces, you have arrived - you can play like you are seeing right through the backs of opponents' cards.

Frequently the opponents are considerate enough to provide what you need to know. They do this by bidding something, or sometimes by not bidding something, while your side is approaching a contract. Their actions can be quite helpful. You should love their bids (when your side has the balance of power of course); and give them a nice thank-you smile when they do a takeout double or a preemptive call; they are so descriptive of honors and distributions.

You can add more data bytes during the play of the hand; especially from the opening lead which is so enlightening. When you still need more info, play the cards in a sequence that enables you to discover the critical missing bytes.

Preemptive calls and overcalls behind notrump bids signify long suits, and at the same time tell you quite a lot about the missing honors. From them you should find the way to win uncertain tricks. It may surprise you how often this thought process turns a likely loss into a win.

* * * *

WORLD TEAM OLYMPIAD - 2004. This deal was played in the round robin; Netherlands vs. Hong Cong.

Board 18
Dealer East
North-South Vulnerable

NORTH
♠ A J 8 5 2
♥ A 8 7 5
♦ 9 7 2
♣ 5

WEST
♠ - -
♥ K 10 4 3 2
♦ 10 5 3
♣ Q 10 7 4 2

EAST
♠ Q 10 7 6 4 3
♥ Q J
♦ K Q 6
♣ J 8

SOUTH
♠ K 9
♥ 9 6
♦ A J 8 4
♣ A K 9 6 3

West	North	East	South
		1♠	1NT
Pass	2♣	Pass	2♦
Pass	3NT ///		

East opened with a hand-full of "quacks". South (Jansma of the Netherlands) overcalled 1NT with an unbalanced hand of mostly minors, but at least he had a stopper in spades. North tried Stayman then went to 3NT after a negative response. This was a 24 high-card game try, both hands holding good five-card suits.

West led the ♣4 to the ♣5, ♣J and ♣K. It was obvious to Jansma that East had most of the outstanding honors to support his opening bid, and that West led from length. Jansma had six top tricks, needing three more. He set out to discover, playing the ♠K, West discarding a heart.

Now Jansma could visualize nearly the entire defense holdings. East started with nearly all of the honors except maybe a ♣Q and ♥Q, and six spades. If, as is likely, West led from five clubs, then East had two clubs and five red-suit cards.

Jansma played a low heart, ducked to East's ♥J, and won East's club return. He then played a heart to his ♥A, noting East's ♥J falling. Now he led a diamond, allowing East's ♦Q to hold; and East had to return diamonds. Jansma ran the remaining diamonds then put East in with a spade. East had no choice but to return a spade to the dummy.

This was a hefty win. In the other room Hong Cong played 3NT doubled; down two and minus 15 IMPs.

* * * *

WORLD TEAM OLYMPIAD - 2004. In the women's division, Russia (where bridge players were less than 1% of the number in the United States) made it to the finals against the top USA team. Seemingly improbable, Russia won 271 to 259.

Board 25
Dealer East
North-South Vulnerable

NORTH
♠ J 7 4
♥ A J 2
♦ K J 9 5 4 3
♣ A

SOUTH
♠ K 6 2
♥ Q 9 4
♦ Q 7
♣ Q 10 5 4 3

West	North	East	South
		Pass	Pass
1♠	2♦	Pass	3♦
Pass	3♠	Dbl	3NT ///

USA managed to find this thin game after Russia opened 1♠. West led the ♠3. Declarer (Montin) saw five diamonds, one club, two hearts if the king was onside, and a spade; sufficient for the game. But one spade stopper would not suffice because she had to concede the ♦A in order to establish her diamond suit.

Montin faced a tough decision. If West had both spade honors, the ♠J would be best from dummy. On the other hand if East had a spade honor, the better play would be low from dummy hoping the ♠J would become a second stopper. Montin decided that East's double likely promised a spade honor.

The full deal was:

```
Board 25                      NORTH
Dealer East                   ♠ J 7 4
North-South Vulnerable        ♥ A J 2
                              ♦ K J 9 5 4 3
                              ♣ A
          WEST                                EAST
          ♠ A 10 8 3 2                        ♠ Q 9
          ♥ K 10 8 5                          ♥ 7 6 3
          ♦ A 8 2                             ♦ 10 6
          ♣ 8                                 ♣ K J 9 7 6 2
                              SOUTH
                              ♠ K 6 2
                              ♥ Q 9 4
                              ♦ Q 7
                              ♣ Q 10 5 4 3
```

At the first trick, Montin played low from dummy and won in hand: ♠3, ♠4,♠9, ♠K. The ♠9 was a good play for East but it didn't matter. When West later won the ♦A and continued spades, East won the ♠Q but could not continue the suit.

*   *   *   *

There has been a noticeable trend toward light openings in tournaments. Of course most often you end up on the defense after your light opening. It follows that whenever the opponents win the auction, they will have gained information, sometimes crucial information, about your holdings, hence your partner's holdings.

Every defensive or interfering bid you make, you reveal key information to the other side.

Here we have focused on locating defense honor cards. The information-gathering methods used to do this are the very same ones employed to fill in empty spaces. They are worth repeating briefly.

> An opponent's bidding tells you everything it tells his partner.
>
> Regarding the opening lead, generally it is evident if it is attacking or passive, then easy to determine the profile of the suit led. Well, it is not always easy; your opponents may have systemic lead rules that could make a difference.
>
> As to both bidding and opening leads, there is a wealth of additional information to be gleaned by inference, in particular, from what was not bid and what was not led. I suggest there is often more to learn from these negative, non-barking, non-events than from actual actions.
>
> Finally, when you still need more bits of information, there is the discovery process, playing non-critical tricks to further reduce the unknowns and nudge the odds further your way.

10
YE OLDE QUANDARY

The controversy about majors vs. notrump has survived for decades. It's all about hands that qualify to open a five-card major (1♥ or 1♠) and also 1NT. These are hands valued at 15 to 17 high-card points (or 16 to 18 if you prefer) and contain a five-card major. This overlap in opening requirements exist because distributions of 5-3-3-2 are considered sufficiently balanced so as to qualify for a notrump opening. When the five-card suit is a minor, there seems to be little or no debate; just about everyone opens 1NT. But when the five-card suit is a major, we have had more controversy and less clarity.

Some bridge players go to considerable lengths to avoid notrump contracts; but the rest of us, because we are driven to win, search for the best contracts whatever the strain. Heretofore we have not known what the best practice is, even if we thought we did.

This is an old debate, going back to the introduction of five-card majors, if you can remember that long ago. Since that olden time, there have been lots written about this quandary. Here's how several gurus weighed in on it:

> In ***Common Sense Bidding*** William Root's view was: "It is often right to open the bidding with notrump when holding a five-card major suit (especially a five-card heart suit) to avoid rebidding problems. However, with a very strong five-card major and/or a worthless doubleton, usually it is better to bid the major suit."
>
> Some years later, Marty Bergen in his book ***Points Schmoints*** wrote: "Whenever you have a balanced hand and the appropriate point count, open 1NT. There are absolutely no exceptions. Do not be distracted by a five-card major."
>
> Contrastingly we have this comment by a notable guest speaker at an ABTA Convention some years ago: "I always open a five-card major, even when the hand satisfies notrump requirements." In all fairness, I must point out that this speaker only expressed his preference; he did not say majors are the better choice.

Does this makes you wonder? Perhaps the choice is so close that it doesn't matter much over the long run; but then maybe it does. And if it does, then I want to know the best practice, if there is one, and you should too.

For starters, take a look at these two hands.

♠ A Q J 7 2
♥ J 9
♦ Q 3 2
♣ A K 3

After you open 1♠, what do you then do when partner says 2♣, 2♦ or 2♥? Personally I prefer not to rebid spades with only five; besides that would understate the strength of the hand. I might try 2NT; maybe partner will give me a delayed spade raise, then maybe not. I think this deal should play in game somewhere and notrump looks like a reasonable option, so I might jump directly to 3NT.

The rebid question can be even more troublesome if your suit is hearts and partner returns a spade. Here's a similar hand with hearts and spades reversed.

♠ J 9
♥ A Q J 7 2
♦ Q 3 2
♣ A K 3

If you open 1♥ and partner returns 1♠, what will you do next? You can't support spades. A heart rebid isn't the worst choice, but partner will no doubt figure you for six hearts. Alternatively 1NT understates your values and 2NT overstates them.

These are real concerns about majors. Bergen avoids them by opening 1NT, immediately defining his strength and distribution with some precision while concealing his five-card major. After 1NT, the partnership has no convenient way to find a 5-3 major if there is one; thus is resigned to playing 3NT instead of 4♥ or 4♠ when the partnership has the 5-3 major fit and game values. (If you are inclined to say "so what", hang in - you will find clarity anon).

To be sure, sometimes 3NT will produce a higher score; but then other times 4♥ or 4♠ will produce better results. What is not clear is which approach produces the better results more often; or in the event there are hand characteristics, other than distribution, that determine the better choice, what these defining characteristics are. Certainly I was skeptical about Bergen's unqualified stance in favor of notrump so I set out to find reality, being entirely open of course toward being convinced either way.

When partner opens a five-card major and you have three or more, you know immediately that you have a good suit in which to play the contract. Usually this scenario produces straightforward bidding sequences. However when you do not have support, opener may have difficulty describing his strength, particularly when he has more than a minimum and not quite enough for a jump shift (hands of 15 to 18 points).

Alternatively, when you open 1NT, partner is immediately aware of your strength but lacks knowledge of your five-card major. It would appear that the better choice depends upon the character of partner's hand, which you would like to know but don't, that is if you are playing by the rules.

The first step in coming to grips with this quandary was to segment these deals into three subsets or categories, each specifying the amount of support partner has for the strong opener's five-card major. The three categories are:

Super Fits: Deals where the responder has at least four-card support for partner's five-card major.

Misfits: Deals where the responder has less than three-card support for partner's five-card major.

Golden fits: Deals where the responder has exactly three-card support for partner's five-card major.

Even though you don't know which of these conditions are present when you pick up your good hand and contemplate an opening gambit, understanding the implications of each is an essential prelude to sorting out the issues and defining best practice.

Super Fits

If a 5-3 fit is "golden", then a 5-4 or 5-5 fit must be more so. We could dub them "plutonium" or "oil", but that might confuse focused bridge players, so I refer to them simply as "super fits". No matter how you bid them, you usually find a good contract; and quite often they are a piece-of-cake to play - just pull the few outstanding trump early and ruff away losers.

When your partnership has more than enough trump you can easily find these super fits, whether opening the major or notrump, because you have the tools created by Sam Stayman and Oswald Jacoby that enable you to find good major fits following notrump openers.

Yes, you can always find these fits, but there are some particular deals where you have no need, or should not want, to find them. Recall from Chapter 2 - FLATS that you do better at duplicate (six out of ten deals) electing to play notrump rather than the major given a 5-4 fit and a flat responding hand. Sure the choice is close; the edge favoring notrump is mostly due to the 10-point margin earned for the first trick. Excluding flats, major suit contracts average better results when either hand is unbalanced. With this understanding, your objective when there is a 5-4 fit should be to play in notrump when responder's hand is flat, and to play in the major otherwise.

How do you achieve this simple objective? Consider these three hands after partner opens 1NT.

A)	B)	C)
♠ K Q 4	♠ K 3	♠ J 6 4
♥ Q J 6 2	♥ J 10 8 7	♥ J 7 5 4 3
♦ A 10 5	♦ A 8 5	♦ A 5
♣ 8 5 4	♣ J 8 7 6	♣ A Q 10

Hand A: Flat but has game values. The best plan is to ignore a possible 4-4 or 4-5 heart fit and raise directly to 3NT.

Hand B: Nearly balanced with invitational values and four hearts. I suggest you test hearts before settling on notrump: bid 2♣ (Stayman) to check for a heart match; if partner returns 2♥, invite to game with 3♥; otherwise bid 2NT.

Hand C: The common 5-3-3-2 profile with game values. Use the transfer: call 2♦, alerting partner that you have five hearts; then continue to 3NT. You have given partner the information he needs to chose hearts or else pass 3NT.

You can steer the bidding as you wish following a 1NT opening, doing this by way of the selective use of the Stayman and Jacoby conventions. Additionally you always find 5-5 fits along the way via the transfer mechanism; and of course you are pleased to play those 5-5 fits in the major.

Having considered how to handle three hands following a notrump opener, alternatively what if partner opens his major? Let's assume that partner opens 1♥ instead of 1NT. How do you now respond holding the same three hands.

A)	B)	C)
♠ K Q 4	♠ K 3	♠ J 6 4
♥ Q J 6 2	♥ J 10 8 7	♥ J 7 5 4 3
♦ A 10 5	♦ A 8 5	♦ A 5
♣ 8 5 4	♣ J 8 7 6	♣ A Q 10

Hand A: As soon as partner opens one heart, you know there is a 5-4 heart fit and game values. While you have a flat hand, you do not know the shape of partner's hand and have no convenient way to discover it. Remember my bias toward notrump, and hopefully yours too, involves flat hands opposite balanced openers. Since responder does not know the shape of opener's hand, the only recourse is to proceed to the heart game. ("Only recourse" is not meant negatively. There cannot be very much wrong with a 4♥ or 4♠ contract holding game values and a 5-4 trump suit.)

Hand B: Since this hand has a doubleton, when partner opens 1♥ there is no reason to be interested in anything other than a heart contract. Raise to 2♥ or 3♥ depending on your style. You should reach 4♥ whenever partner has the requisite strength.

Hand C: With a known 5-5 fit and your unbalanced values, you should easily reach a 4♥ game.

Regarding these alternate approaches, you should achieve the most desirable contract with hands B and C no matter which opener partner elects. Hand A is problematic; most pairs will end in whatever strain, hearts or notrump, partner opens. If the major is selected, you will have conceded the 10 point advantage notrump so often has over the majors; a substantial concession to be sure in match-point bridge.

In fairness I should point out that there is a way to find the better notrump hand with certain flats such as hand A above. For example:

WEST	**EAST**
1♥	3♥
3NT	

After a heart opening, East jump-raises to show support and invitational values. West rebids 3NT in spite of the 5-3 heart fit to tell partner she has a balanced hand and to give her an option of hearts or notrump. This hand would be a good candidate for West's 3NT call:

♠ Q J 8 ♥ A J 10 5 4 ♦ K 10 ♣ K Q 8

* * * *

WORLD TEAM OLYMPIAD - 1992. USA took on France for the gold medal. The Americans were defending the title they had won four years earlier. Here was the first of 96 boards played.

Board 1
Dealer: North
None Vulnerable

NORTH
♠ A K 3
♥ A 10 6 3
♦ A 8 7 4
♣ 8 2

WEST
♠ 8 6 5 4
♥ J 8 2
♦ 9 6 2
♣ A 9 3

EAST
♠ Q J 9
♥ 9
♦ Q J 5 3
♣ K Q J 6 5

SOUTH
♠ 10 7 2
♥ K Q 7 5 4
♦ K 10
♣ 10 7 4

North (Deutsch USA) opened 1NT and Rosenberg, with a balanced 5-3-3-2 hand, raised to 3NT. Lady Luck would have her way, in this instance for the defense. East naturally led a club, and the defenders ran five straight tricks.

In the other room, North (France) also opened 1NT, but South (Mouiel) transferred to hearts which got them to 4♥ with a 5-4 trump fit. This game was cold as declarer needed only ruff one club in his hand for the 10th trick. The super fit won out, as it mostly does when neither hand is flat. France jumped ahead 10 IMPs.

* * * *

When your partnership has a super fit, you are able to find the good major following either opener - major or notrump. However, pairs who open the major are at a disadvantage when responder has four-card support and a flat hand as neither player has sufficient knowledge to divert the bidding from the major into the superior notrump. Responder does not know his partner has a 5-3-3-2 distribution; ***and that is the critical factor giving notrump openers a ten-point edge.***

Neither approach has an advantage when the fit is 5-5 since the better strain can be readily found no matter which opening is chosen.

Misfits

Chaos! You have good opening values plus a five-card major. What more could you wish for? Well, maybe first you might wish that partner has support for your major. Otherwise you run the gamut of misfits and other undesirables that you neither wish for nor relish: partner loves his suit, you hate it; one opponent has five of yours, the other has none; finesses are offside, nothing breaks. These evils seem to happen often when your hands misfit, even mildly. You cannot avoid them nor ignore them. The opponents hold lots of your suit, and they hold them about 46 deals out of every 100 where you have the choice of opening a major or notrump.

It would seem to be self evident that, if you knew in advance which deals partner had no support, you would certainly open 1NT rather than the major. After all what would be the point of the major when partner only had one or two of them, maybe none, while the enemy had six, seven or eight. Surely the bidding would be easier not having to waste space and mental anguish on a suit of less than eight. On the other hand what matters if the bidding is sometimes difficult or awkward if you find the best contract. But then perhaps the best sequences are actually the simplest ones.

* * * *

Generally it is advantageous for the strong hand to become declarer, no matter the strain; so much so, that Jacoby Transfers were born. It does not always matter, but you certainly want to be in the best hand when it does matter.

Dealer: South
Vulnerable: N-S

NORTH
♠ 10 3
♥ K 9 8 4
♦ Q 5 3
♣ Q J 5 4

WEST
♠ K 9 8 6
♥ J 10
♦ K 7 4
♣ 10 7 6 2

EAST
♠ 5 4
♥ A 7 6 5
♦ A 10 8 6 2
♣ 9 8

SOUTH
♠ A Q J 7 2
♥ Q 3 2
♦ J 9
♣ A K 3

WEST	NORTH	EAST	SOUTH
			1NT
P	2♣	P	2♠
P	2NT	P	3NT ///

Undeterred by his meager diamonds, South opens 1NT. North tries to find a fit in hearts then calls 2NT and North accepts the invitation. Avoiding the majors, West leads his fourth club. This is a secure contract (four spades, four clubs and one heart) so long as South stays away from diamonds.

If South opens 1♠, the bidding is not so sure-footed.

WEST	**NORTH**	**EAST**	**SOUTH**
			1♠
P	1NT	P	2NT
P	3NT///		

With 17 high-card points, South presses on to 2NT. North has but eight high-card points; well, it's better than a minimum so why not try for game!

In this sequence North becomes declarer, a common outcome when the strong hand opens the major. East leads his fourth diamond.

Declarer perceptively plays the dummy ♦9 which enables him to win the third diamond with the ♦Q. This, however, only lessens the bleeding. The defense wins four diamonds, one heart and one spade, down two. Chalk up a big gain for the notrump openers: 600 for making compared to -200.

*   *   *   *

Dealer: South
East-West Vulnerable

NORTH
♠ 10
♥ A Q 10 6 3
♦ Q 10 8 2
♣ J 8 3

WEST
♠ A 6 5 2
♥ J 9 8
♦ J 7 3
♣ Q 10 4

EAST
♠ J 9 3
♥ 5 4
♦ A 6 5 4
♣ K 6 5 2

SOUTH
♠ K Q 8 7 4
♥ K 7 2
♦ K 9
♣ A 9 7

WEST	**NORTH**	**EAST**	**SOUTH**
			1NT
P	2♦	P	2♥
P	3NT	P	4♥///

Following the 1NT opener, North uses the transfer to promise hearts then aggressively jumps to 3NT. With good heart support, South dutifully converts to the heart game. West leads the ♣4 to the ♣2, ♣K, ♣A. Fortunately this provides a second club trick to declarer. By successfully finessing the ♦J, declarer loses only three tricks - one club, one diamond and one spade. Here the transfer works efficiently in conjunction with the notrump opener.

Suppose South opens 1♠. With North's holding, most of us would respond 2♥. The meek might end in a part score contract, but most competitors would charge ahead to 4♥. In this instance the results would be the same.

*   *   *   *

Here are three of a near-infinite number of non-fit hands you might pick up opposite a strong partner opening 1NT:

A)	B)	C)
♠ 9 2	♠ J 10 8 2	♠ K Q J 9 2
♥ 8 6	♥ 10 9	♥ - -
♦ 9 6 2	♦ 9 6 5 3 2	♦ Q J 10 9 8 2
♣ K Q 9 8 6 5	♣ A 7	♣ 10 4

Hand A: Surely this hand will play well in clubs. The usual way to bid this when partner opens 1NT is to call 2♠ as a transfer to 3♣ and pass. Instead maybe you would jump directly to 3NT; in which case you would be in the company of the tigers.

Hand B: No bid; you have very little to offer. This hand is not worth looking for a 4-4 spade fit.

Hand C: Common practice is to transfer to spades then make a free bid in diamonds at the three-level.

* * * *

Let's look at the consequences with these same hands when partner open 1♥ instead of 1NT.

A)	B)	C)
♠ 9 2	♠ J 10 8 2	♠ K Q J 9 2
♥ 8 6	♥ 10 9	♥ - -
♦ 9 6 2	♦ 9 6 5 3 2	♦ Q J 10 9 8 2
♣ K Q 9 8 6 5	♣ A 7	♣ 10 4

Hand A: This hand is too weak to do anything but response 1NT to a 1♥ opener. Unless your 1NT is forcing, partner will pass. This 1NT contract will almost certainly be inferior to 3♣, but only because you have a good six-card suit.

Hand B: You have just five high-card points; partner opens 1♥. You could stretch a bit and respond 1♠, but you never get the chance; right-hand opponent overcalls 2♣ and you are relieved of a decision. Shucks, 2♣ makes for East-West as does 1NT North-South. Such is the preemptive value of opening 1NT.

Hand C: Partner opens hearts, your void. This does not look like a notrump. How to describe both suits - do you start with the longer minor or the shorter major? Maybe you could start with 1♠ followed by bidding diamonds twice - if the bidding doesn‘t get too high.

* * * *

When you open the major, being unable to support it, partner quite often returns 1NT; and with some regularity notrump then becomes the contract. This sequence results in the responder becoming declarer and the strong hand being tabled as dummy. Alternatively opening 1NT keeps the strong hand protected whenever (quite often), the bidding ends in a notrump contract. This notion, that it is advantageous to have the strong hand become declarer is widely held. While it is not the huge advantage some pundits would have you believe, it does pay off more often than not.

When responder's hand is weak, the odds favor playing in his five-plus suit at the two-level in preference to one notrump. This of course refers to responder hands that are too weak to go beyond the one or two level. When his suit is a major, he should transfer and pass. Notrump openers have this small, but important advantage because otherwise (opening an unsupportable spade suit), responder must pass or call 1NT as his hand is too weak to do anything else.

When you have one of these hands that can be opened either a major or 1NT, 46% of the time partner does not have support for your major, not much less than one-half of the time. Of this universe of misfits, notrump openers win 4 of 20 deals, whereas suit openers win 1 of 20 deals. The remaining 15 produce, or should, the same contract after either opening gambit.

Notrump is the clear winner when partner cannot support your five-card major.

Golden Fits

Bridge lore has it that whenever a partnership has a 5-3 fit in a major, the contract should be played there. To support this notion we have good bidding procedures to find golden fits and to determine exactly how high to bid them. In olden times, we favored the major over notrump, hoping to find the golden fit. Statistically we find three-card support in responder's hand 29% of the time.

When partner opens a major and we have three-card support, mostly the contract is played in the major. But, surely we would like to play some of them in notrump; deals where our hand is flat and partner's hand is 5-3-3-2, because, as was pointed out earlier (Chapter 2: FLATS), such deals consistently produce better results in notrump. We are unable to do this after a major opening because we have no way to determine that partner's hand is balanced. Short of creating a new convention to uncover this characteristic, we cannot steer the bidding into notrump. Hence (extreme distributional deals aside) when partner opens a major and we have three of them, these deals will be played in the major.

Alternatively, when partner opens 1NT and between us there is a 5-3 major, we have no way to uncover this information; neither Stayman nor transfers are suitable. Consequently they are played in notrump.

* * * *

BERMUDA BOWL - 1999. This world championship was nearly owned by the USA teams. USA1 won the championship by a score of nearly 2 to 1 over Brazil, and USA2 came in third. Earlier during the round robin eliminations there was this deal exactly of the kind we are considering. North had a 15 point balanced hand and the partnership had a 5-3 spade fit. In France vs. Norway, the deal was played both ways (3NT and 4 spades) with results we can now predict.

Board 13
Dealer: North
Both Vulnerable

	NORTH	
	♠ A Q J 9 3	
	♥ K J 6	
	♦ J 6	
	♣ K 10 5	
WEST		EAST
♠ 8 7 6 5		♠ 4
♥ A 8 2		♥ 9 5 3
♦ Q 8 7 2		♦ K 10 5 3
♣ 9 7		♣ A J 6 3 2
	SOUTH	
	♠ K 10 2	
	♥ Q 10 7 4	
	♦ A 9 4	
	♣ Q 6 4	

WEST	NORTH	EAST	SOUTH
	1NT	Pass	3NT ///

Norway playing North-South bid notrump directly to game without pause to consider spades. East led off with his fourth club, allowing declarer (Saelensminde) time to drive out the ♥A and bring in eleven tricks. A diamond lead could have doomed the contract, but East had no clue.

In the other room where France (Multon/Mari) held the North-South hands, the bidding began with spades, and unsurprisingly ended in 4♠.

WEST	NORTH	EAST	SOUTH
	1♠	Pass	2♣
Pass	2♠	Pass	3♠
Pass	4♠///		

This contract looked solid except for the 4-1 trump split. Regardless, Multon adroitly brought home the contract. However notrump won by 40 points, just 1 IMP in this team contest.

* * * *

WORLD TEAM OLYMPIAD - 1992. It was early in the contest between USA and France for the championship. In the French, we see a different style of bidding.

Board 9 (Rotated)
Dealer: East
North-South Vulnerable

NORTH
♠ 9 8 4
♥ J 8 4
♦ A K J 8 5
♣ J 9

WEST
♠ A 7 2
♥ K Q 9 3
♦ 3 2
♣ 10 7 6 3

EAST
♠ J 6
♥ 10 7 6 2
♦ 10 7 4
♣ A Q 8 2

SOUTH
♠ K Q 10 5 3
♥ A 5
♦ Q 9 6
♣ K 5 4

WEST	NORTH	EAST	SOUTH
USA	France	USA	France
		Pass	1♠
Pass	2♦	Pass	2♠
Pass	3♠	Pass	4♠///

It was aggressive bidding by the French pair. Declarer lost the trump ace, one heart and one club; ruffing a club in dummy for his 10th trick. Result: Game and +620 for France.

In the other room USA's Rodwell (South) chose to open 1NT rather than spades. Confirming that the French had no monopoly on aggressiveness, Meckstroth jumped directly to 3NT.

West led the ♥Q, producing a second stopper for Rodwell, giving him time to establish his spade suit. Result: Eleven tricks and +660.

*   *   *   *

VENICE CUP - 2007. USA and French teams were in the running for the women's championship, and met head-to-head in the quarterfinals.

Board 81
Dealer: North
None Vulnerable

NORTH
♠ 9 6 2
♥ K 7 5
♦ 8 4 3
♣ K Q 6 2

WEST
♠ K J 8 5
♥ 4
♦ A 10 9 5 2
♣ 8 7 4

EAST
♠ 10
♥ A 10 9 8 6 3 2
♦ K 6
♣ 10 5 3

SOUTH
♠ A Q 7 4 3
♥ Q J
♦ Q J 7
♣ A J 9

WEST	NORTH	EAST	SOUTH
	Pass	3♥	Dbl
Pass	3NT///		

This was a by-the-book preempt by East (USA2). South might have overcalled 3♠ with this attractive opener, but instead chose the takeout. Needing to bid and having a heart stopper, North (D'Ovidio, France) went for the notrump game.

The ♥10 was led by East. This rated to be difficult for declarer (North). The good news was that West almost surely had only one heart, while the bad news was that West had most of the spades behind dummy and D'Ovidio needed a diamond and two spade tricks. A low spade was played, West's ♠J winning. Then it was a diamond to East, and the ♥A and another heart to declarer. Declarer ran four clubs, then a diamond to West, who was end-played into leading a spade for North's ninth trick and game.

In the other room, East also preempted but South (USA2) overcalled spades rather than doubling.

WEST	NORTH	EAST	SOUTH
	Pass	3♥	3♠///

Sandborn (USA2) playing 3♠ had no chance. The defenders won the ♥A and a ruff, two top diamonds and two trump tricks for down 2 and 11 IMPs to France.

*   *   *   *

The risk of the defense running a suit is often present in notrump contracts, although it doesn't happen as often as you may expect.

Dealer: South
Vulnerable: N-S

NORTH
♠ 7 6 4
♥ Q J 5 2
♦ A K 7 6
♣ 10 6

WEST
♠ A 10 9
♥ 10 9
♦ 5 3 2
♣ Q 9 8 7 3

EAST
♠ 8 3
♥ 7 6 4 3
♦ J 10 8
♣ A K 4 2

SOUTH
♠ K Q J 5 2
♥ A K 8
♦ Q 9 4
♣ J 5

At table 2 South opens 1♠ and the pair continues to 4♠. Declarer loses two clubs (assuming West finds a club lead at the outset or when he gets in with the trump ace) and the ♠A, making an easy game contract, for +620.

At table 1, South opens 1NT and North raises to 3NT. West leads the ♣7 and declarer loses five clubs quickly then later the spade ace; down two vulnerable for -500.

These results produce a huge swing in any bridge contest; a top or bottom in match-points, depending on which side of the fence you are; 1120 points in rubber; 15 IMPs in team contests.

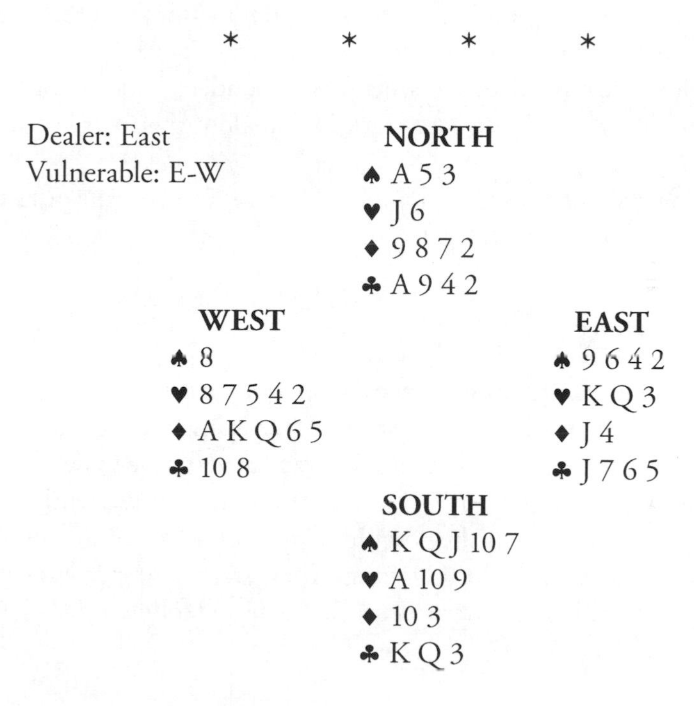

Table 2: South opens 1♠, North raises, South invites to game, and North, with better than a minimum, accepts. In the play, declarer has nine quick tricks and finesses hearts (East must cover the ♥J) for the 10th trick and game.

Table 1: South opens 1NT and North raises directly to 3NT. West leads the ♦K. The consequence of that lead is that the defense is now unable to run diamonds. If West continues diamonds, he will promote the ♦9 for declarer. However the hand is played out, declarer wins five spades, three clubs, one heart, and either another heart or diamond, for ten tricks.

The notrump players win this deal by a mere 10 points, an outcome we see quite frequently.

*   *   *   *

Assume South is dealt a 1NT hand with a five-card major, and North is dealt any random hand that includes exactly three of South's major. In all of these deals North would find a flat hand in 31 of 100. When North gets one of these flat hands, the edge is about 2:1 in favor of notrump. Thus almost 7 of 10 such deals produce better scores in notrump; and this edge is achieved by opening in notrump rather than the major.

Expectations are quite different when the responder's hand is not flat. Then notrump has a small win/lose edge, but so close that practically it is a draw.

Combining the outcomes of all golden fits give us a margin in favor of notrump of about 55-45.

Winning Strategy

We have pondered the consequences of opening in a major compared to 1NT. Responder hands were categorized according to the extent of support for the opener's five-card major. In practice you have to decide on the initial call with no knowledge of partner's hand. Consequently, the best overall results depend on the combined odds of all three categories. Putting them together yields the following. These statistics represent win/loss rates comparing likely best contract scores achieved following 1NT vs. five-card major openers.

	MAJORS	**NOTRUMP**	**TIES**
SUPER FITS	4	7	14
MISFITS	2	9	35
GOLDEN FITS	12	15	2
	18	31	51

And so returning to "ye olde quandary" we see that Bergen's unqualified stance favoring notrump openers is absolutely correct.

11

HIDE & SEEK

Declarers do not have proprietary rights to bidding clues. While defenders spend most of their time playing in shadows, any clues you give them allows some light to filter in. So it is good tactics to keep them in the dark if you can by whatever means.

Perhaps you should be more frugal in your bidding; avoiding calls that do not advance your partnership toward better contracts. This is fairly easy when the likely strain is notrump, and more so when your length is in the minors.

* * * *

Dealer West
East-West Vulnerable

WEST	EAST
♠ K J	♠ 7 6 2
♥ Q J 6 4	♥ A K 5
♦ A Q 9 7 6	♦ 5 4
♣ A J	♣ K 10 8 7 3

There are three common ways to bid these two hands. Presumably you end in 3NT one way or another, because the prospect of a notrump game is quite good.

The point of this exercise is to think about how different bidding styles might influence defender play for better or worse.

1)

West	North	East	South
1♦	Pass	2♣	Pass
2♥	Pass	3NT ///	

West describes his hand as five diamonds and four hearts, with 16 to 18 points. Lacking a spade stopper, East gambles a bit and bids the notrump game. This pair bid three suits along the way. On lead, South would of course lean toward a spade lead.

2)

West	North	East	South
1♦	Pass	2♣	Pass
3NT///			

West chooses to jump to 3NT after hearing a 2♣ response (which denies a five-card major). This sequence withholds information to the defense about the heart suit at no cost.

3)

West	**North**	**East**	**South**
1♦	Pass	1NT	Pass
2NT	Pass	3NT///	

Rather than show his good minor East responds 1NT, effectively ending consideration of a suit contract. Since minor openers do not promise much length, South on lead has little guidance and likely would lead his best/longest suit, which randomly could be any of the three unbid suits, or perhaps even diamonds.

The first sequence is the one most likely to elicit a spade lead. If the defenders take four spades, the contract is doomed. The third sequence is the least informative, and so has the best chance of avoiding a spade lead.

It is self-evident that the goal of bidding is to find the best contract. It should be equally self-evident that bids that are not essential to that goal, not matter how descriptive they may be, should be avoided.

*   *   *   *

Dealer West	**WEST**	**EAST**
None Vulnerable	♠ A 2	♠ K Q 6 4
	♥ Q J 8	♥ A 9 3 2
	♦ 6 5	♦ K Q
	♣ A Q J 9 5 2	♣ 10 6 4

The issue here is: where to play the contract - 5♣ or 3NT? Either contract makes when the ♣K is onside. When it is not onside, 5♣ makes when the ♥K is onside. Alternatively 3NT makes whenever the defense does not find a diamond lead.

The club strain makes game more often than 3NT by a small margin, but here is the catch. Every time the ♣K is onside (50%), notrump wins the match-point game because it takes as many tricks as does 5♣, while earning higher scores every time.

If you were way behind in the tournament, you might try 4♥. If you found hearts dividing 3-3, you could win a top board no matter where the ♣K, but don't bet the pot - the odds in your favor are below 20%.

Now, how to bid East's hand? With this nearly flat hand, notrump is likely to be superior to a 4-4 major. Hence I recommend this stingy bidding:

West	**North**	**East**	**South**
1♣	Pass	3NT*///	

* Limited 13-15 HCP

East's bid is descriptive, defining distribution and strength sufficiently for West to evaluate slam prospects; so you should not miss slam Most important, you have not given aid to the opposition.

*   *   *   *

Dealer West None Vulnerable	**WEST** ♠ K J 6 ♥ K Q 7 6 5 ♦ A Q J ♣ 7 2	**EAST** ♠ Q 9 7 3 ♥ 9 8 3 ♦ 7 5 4 ♣ K Q J

West has a meaty hand of sixteen high-card points. Opposite this, East a flat eight points. Here are three alternate bidding approaches, in order from the least to the most desirable, in my judgment.

1)

West	**North**	**East**	**South**
1♥	Pass	2♥	Pass
3♥	Pass	4♥ or Pass///	

East's single raise is a standard weak response. At the second round, East has a close call, his flat 4-3-3-3 being a detraction.

2)

West	**North**	**East**	**South**
1♥	Pass	1NT	Pass
2NT	Pass	3NT///	

Rather than raise hearts or bid 1♠, East responds 1NT because of his shapelessness. The end result of this decision is a notrump game instead of hearts.

3)

West	**North**	**East**	**South**
1NT	Pass	2NT	Pass
3NT///			

West's hand presents the classic choice between a major or 1NT. After a 1NT opener, East has no interest in looking for a 4-4 spade fit so the pair gets to the notrump game.

In the third sequence, East-West gets to the notrump game without giving the defense any information regarding distribution.

*　　*　　*　　*

WORLD TEAM OLYMPIAD - 1992. Here were France and USA competing in the finals. On stage was Wolff and Hamman (North-South) playing against Levy and Mouiel.

Board 90
East Dealer
Both Vulnerable

NORTH
♠ A Q J 2
♥ J 7 5
♦ 7 6
♣ Q 8 5 4

WEST
♠ 7 6 5
♥ A 6 3
♦ 9 8 4
♣ K J 9 7

EAST
♠ K 10 9 4 3
♥ Q 9 8 4 2
♦ 10 5
♣ A

SOUTH
♠ 8
♥ K 10
♦ A K Q J 3 2
♣ 10 6 3 2

West	North	East	South
		Pass	1♦
Pass	1NT	2♦*	3NT///

* Hearts & spades

Hamman opened 1♦ and Wolff, following Hamman's rule, skiped spades in favor of 1NT. Hamman then wasted no time bidding the game. Hamman's rule:

When notrump is a choice among reasonable bids, take it.

East chose his best suit and led the ♠10, taken in dummy. Wolff counted eight tricks including the ♠A, for which he would have no obvious entry after leaving the dummy. At any rate, he led a diamond to his hand, then a club to dummy eight and East's ♣A. Not wanting to lead into dummy spades again, East switched to his other long suit, ♥4, ten, three, dummy jack. That was just what Wolff needed - access to the dummy's ♠A for his ninth trick and game.

A spade was the natural lead for East, but what might he have led if Wolff had bid 1♠ rather than 1NT at his first response? Of course later West might have come up with the ♥A instead of ducking to dummy ♥J. Too many ifs, one certainty: if East had elected to start with a heart, the defense would have prevailed.

* * * *

I do not suggest you blithely steer toward notrump to the exclusion of other potential contracts. A partnership always needs to bid in a manner that insures finding good games and slams. Beyond that however, bidding suits that have neither promise nor help in finding good games or slams is self-defeating as it always enlightens the opponents.

12

SAFE COMBINATION

Most bridge players would concur that Goren's point-count system leaves something to be desired. After all, it's been more than a half century since Goren and his colleges created it. I marvel at how the 4-3-2-1 honor value scale has survived. Goren's system provided a sensitive balance between accuracy and simplicity, which of course is the very reason it has survived, mostly intact.

There have been many modifications to Goren's hand evaluation system proposed over the years, mostly marginal changes, leaving the core intact. A few have stuck, most not.

In 1985 Goren's editors conceded that distribution could contribute added value to notrump hands, but they stopped short of defining or quantifying distributional values.

If you aspire to play superior bridge, you need to better understand how suit length and quality impacts on hand values. No doubt some experts are already are up to snuff on this; but they are not talking about it; they're too busy winning.

Can 25 points produce a notrump game? Yes, of course, sometimes. So can 23 points, but I would rather put my money on the other side of that bet. At 25 points, the odds are not too bad. Here we have a typical (if there is such a thing in bridge) 25-point deal.

NORTH
♠ A 9 8 3
♥ Q J 10 8
♦ A 8
♣ K 10 7

SOUTH
♠ K 7 6
♥ 7 5
♦ Q 9 6 5
♣ A Q 9 6

You are South playing 3NT. West leads ♠Q to the ♠3 - ♠4 - ♠K in your hand. It is likely that West is leading from a sequence, probably Q-J-10 and one or two spots.

This contract looks promising. You can see three spades finessing West for the jack and 10, three and maybe a fourth club, and the diamond ace, for seven or eight tricks. All you need do is drive out two top hearts, finding one honor in West's hand to give you two heart tricks. Along the way West might help by leading into your ♦Q.

*   *   *   *

Here we have another 25-point deal to ponder.

NORTH
♠ A 7 4
♥ 7 3
♦ 8 7 4
♣ A J 10 6 3

SOUTH
♠ K 9 3
♥ A Q 6 4
♦ A K 3 2
♣ 8 5

Again South is playing 3NT. West leads the ♠5. You have six quick tricks, so you need three more clubs or two clubs and the ♥Q. Duck the first spade; then take the second spade in hand with the ♠K. Now finesse clubs twice (returning via a diamond), hoping to find at least one honor in West's hand.

Chances are that East will take the first club and return another spade to retire your last spade stopper. (But then West began with only four spades). Still, if West has one of the two club honors, you should make the contract. Return to hand with a heart to the ace then finesse clubs again. The odds of finding West with at least one club honor are in your favor. Of course it is the five-card club suit that enables this good game try.

* * * *

Experts are not shy about notrump game tries with less than 26 points. If everyone else is in game, you are betting against the field when you stop short. On the other hand, being able to evaluate lite holdings more precisely, so as to sort the stronger from the weaker, will give you a leg up. Actually it's not stronger vs. weaker it's . . . read on.

The source of the revelations you are about to read originate from analysis of world championship play. Some underlying patterns emerged, and they are quite enlightening.

Bidding notrump games with less than 26 high-card points is neither new nor surprising; but the frequency of these actions is unforeseen. Notrump games are bid with considerable regularity where the partnership has less than 26 high-card points, some as low as 22 points. (Actually 22 points is not the lower limit; there is no telling what a desperate pair may try in order to catch up.) These are not highly unusual deals; they are predominantly run-of-the-mill balanced and nearly balanced hands.

Keep in mind that team bridge with IMP scoring favors bidding games when the odds are 50% and perhaps lower. Mike Lawrence in ***Swiss Teams of Four*** advised:

> **At the table you take the view that if game may exist, you should bid it.**

I take Lawrence's advice to mean you should bid games when the odds of making are well below 50%. This results in going down a lot, but then so do your opponents. Perhaps this liberal policy contributes to the numerous thin notrump games played. Nevertheless, with some certainty that there is more going on than just aggressive bidding; I devoted considerable effort to this phenomenon. Are the experts simply doing what Lawrence suggested, or is there something else going on? Most important, if experts actually make a majority of such deals, maybe we all need to update our notrump hand evaluation rules.

* * * *

VENICE CUP - 1991. Well into the final match between USAII and Australia, where USAII tried for a very thin notrump game (23 points) and nary a five-card suit in sight.

Board 69 (Rotated)
East Dealer
East-West Vulnerable

NORTH
♠ A 4 3 2
♥ J 6 4 3
♦ 8 5
♣ A Q 10

WEST
♠ K Q J 7
♥ K 9 7 5
♦ A Q 10
♣ 8 2

EAST
♠ 10 8 6
♥ 10 2
♦ J 7 3 2
♣ J 9 6 5

SOUTH
♠ 9 5
♥ A Q 8
♦ K 9 6 4
♣ K 7 4 3

West	**North**	**East**	**South**
		Pass	1♦
Dbl	ReDbl	Pass	Pass
1♥	Pass	Pass	1NT
Pass	2NT	Pass	3NT///

West doubled South's diamond opener and was forced to bid a four-card major after North redoubled. These actions pinpointed nearly all of the defenders values. North-South continued to a notrump game.

Declarer Passell had only six tricks in sight. Much to her credit, she managed to endplay West into one more trick. At the end, Passell gave the fourth heart trick to West, who then had to return a diamond to Passell's ♦K: down two. This game never had a chance; both hands were well short on honors and neither had shape.

*   *   *   *

BERMUDA BOWL - 1995. It is not uncommon to find Canadian and United States teams reaching world championship playoffs.

Here the Canadians tried for a lite notrump game while USA2 elected to play in a 4-4 spade suit.

Board 20 (Rotated)
South Dealer
Both Vulnerable

NORTH
♠ K 8 5 3
♥ A K 5 2
♦ 10
♣ Q 9 4 2

WEST
♠ - -
♥ Q J 9 8 4 3
♦ J 6 5 4 2
♣ 6 5

EAST
♠ A Q J 10 2
♥ 10
♦ Q 8 7
♣ K 10 8 3

SOUTH
♠ 9 7 6 4
♥ 7 6
♦ A K 9 3
♣ A J 7

West	**North**	**East**	**South**
USA2	**CAN**	**USA2**	**CAN**
			1♦
Pass	1♥	1♠	Pass
Pass	Dbl	Pass	1NT
Pass	3NT///		

South (Mittelman) declared at 3NT. Along the way East (Freeman) overcalled 1♠, potentially a good lead direction, but to no avail as West (Nickell) had none.

On lead Nickell's chance of winning slow tricks in his red suits was remote; if he could establish a suit, he had no entries. So he led the ♣6, which ran to declarer (Mittelman), winning with the ♣7. Mittelman continued clubs with the ♣J, ducked again by East. However Mittelman could not get East to lead a spade; so in the end he had to settle for three clubs, two diamonds, and two hearts. The contract was off two and minus 200 points.

In the other room, USA2 (Hamman) chose the 4-4 spade game. This contract was promptly doubled and set one trick for minus 200 and a tie.

*   *   *   *

WORLD TEAM OLYMPIAD - 1992. Not everyone was bidding lite notrump games, but some were, including USA. North-South had 24 high-card points and near-flat hands.

Board 21 (Rotated)
Dealer: East
Both Vulnerable

NORTH
♠ A 4 3 2
♥ 7 4
♦ 7 6 4
♣ Q J 9 7

WEST
♠ Q J 8
♥ A 10 9 3
♦ K J 9 5 2
♣ 10

EAST
♠ 10 9 7 6
♥ 8 6 2
♦ Q 8
♣ K 8 6 5

SOUTH
♠ K 5
♥ K Q J 5
♦ A 10 3
♣ A 4 3 2

West	North	East	South
		Pass	1♣
Pass	1♦	Pass	1NT
Pass	2♣	Pass	2♥
Pass	2NT	Pass	3NT
Dbl ///			

In USA vs. Sweden, South (Rodwell) began with a strong 1♣ opener and continued aggressively to a notrump game, and was promptly doubled by a skeptical Swede.

On a diamond lead, Rodwell held his lone stopper until the third round. He played a heart to West's ♥A; who then cashed his remaining diamonds and exited with the ♠Q. Rodwell could see five more sure tricks. When he led the ♣Q, East covered with the king, Rodwell's ace won and the singleton ♣10 fell. This gave him one more trick (♣9) for a total of eight. Down one doubled for -100.

In the other room, a more conservative Swede pair stopped at 2NT. This could have been a sizable swing for Sweden, but alas they managed one less trick than Rodwell, for down one and a mere 2 IMP gain.

In the other semi-final contest, France vs. Netherlands, both tables bid and made 1NT. This was a prudent result as declarer was sure only of seven tricks.

* * * *

We reviewed three deals of which none made a notrump game. All of them had two common traits: less than 25 high-card points; and no suit longer than four.

Selecting three hands such as these is an easy exercise when you can choose from hundreds of deals. In fact it was easier than you might think because three out of every four deals having these characteristics did not make game.

It would be nice, and profitable, if you somehow knew which of those deals were good game tries, and bid only them to game. My money says you can't.

If you insist on bidding lite, flat deals to game, you are certain to be a favorite opponent of champions.

So far we looked at lite notrump deals with flat or nearly flat hands, and not coincidentally none made game. Even so, there are oodles of deals that experts bid to game with less than 26 high-card points, a goodly share of them making.

* * * *

BERMUDA BOWL - 1987. On the way to the trophy, the legendary Wolff/Hamman duo played this lite notrump deal.

North Dealer
Both Vulnerable

	NORTH	
	♠ Q J 2	
	♥ Q 4	
	♦ A Q J 8	
	♣ 10 9 3 2	
WEST		**EAST**
♠ A 5 3		♠ 9 7 6 4
♥ 8 6 2		♥ K J 9 7 3
♦ K 10 9 7 3		♦ 6 2
♣ Q 5		♣ K 4
	SOUTH	
	♠ K 10 8	
	♥ A 10 5	
	♦ 5 4	
	♣ A J 8 7 6	

West	**North**	**East**	**South**
	1NT	Pass	3NT///

Hamman opened 1NT and Wolff jumped directly to 3NT, both holding 12 high-card points. Aggressive, surely, but success validates the enterprise.

There was no effective defense. A heart lead to Hamman's ♥Q created a second heart stopper. Hamman set up his long clubs, and finding both the diamond and club kings onside, he lost only two tricks, the ♣Q and ♠A.

*   *   *   *

Should we just bid lite notrump games and rely on our superior play skills to carry us to the winners circle? Searching for some universal truth, I reviewed 744 boards from championship play, examining every deal where someone played a notrump contract.

The first revelation was that there is a large measure of validity in bidding lite deals to game. Overall, experts made game 57% of the time. To the average player this may not sound so great, but in team competition any time you bid games at the margin and make 57% of them, you are winning.

While a 57% win rate for deals at the margin (close calls) may be sufficient in team contests, you should aim higher in match-point contests where you need somewhat better margins to come out on top. At any rate, the issue is moot because you do not need to settle for a 57% win rate.

I found that deals containing less than 24 high-card points made nine tricks so rarely that there is no real hope for them. Excluding hands with long running suits (i.e. gambling notrump), I urge you not to try for notrump games with these extremely lite deals - unless you are my opponent.

Perhaps there is one exception: when you are so far behind that you need a miracle to win, a long shot may be better than conceding and going home early.

Returning to deals of 24 high-card points, the objective was to find some discrete, specific characteristics whereby we can sort out winners from losers in advance.

The initial set of deals consisted of thirty-three boards from several world championships. After careful study, it became apparent that holding a good five-card suit was a decided advantage. Consequently these boards were sorted into two groups. Group #1 contained deals where one hand contained a five-card or longer suit with a minimum of three honor points in that suit. (labeled "qualifying suits"). The remaining deals were placed into Group #2 where a few contained five- or six-card suits without three honor points; and most were balanced with nothing longer than four cards. The statistics for the two groups, where the two hands contain exactly 24 high-card points, are nothing short of astounding. Twenty-four-point deals containing a qualifying suit made games 30 out of 39 times. Absent a qualifying suit they made game a mere 4 out of 19 times. To verify these findings, a second test was undertaken: a reality check if you please. This test duplicated the first using a new data base of 512 deals. Here are these findings.

Deals containing twenty-four high-card points and a qualifying five-card suit made nine tricks 68% of the time. This is less than 77% found in the first trial, but impressive nonetheless. On the flip side, only 10% of non-qualifying deals made nine tricks. This is a powerful confirmation that it is futile to try for notrump games with 24 high-card points if you do not have a qualifying suit.

The presence or absence of a qualifying suit is a highly accurate marker of notrump game potential, one that is far more accurate than anything else in sight. Balanced 24-point deals containing a qualifying suit should be bid to game no matter what contest you are playing - teams, match-point or rubber. They are winners. My short hand label for them is 05-03-24 (a five-card suit - three honors - twenty-four high-card points in the combined hands). This is the safe combination that unlocks a treasure chest of good notrump game contracts.

A deal may qualify for 05-03-24 but that does not mean it will be easy to play. To the contrary, often these deals present difficulties that are quite challenging; they may even be beyond your skill. Some are so problematic that they require help from the defense.

* * * *

WORLD BRIDGE CHAMPIONSHIPS - 1990. This was the first championship event organized entirely by the World Bridge Federation. There are times when you need a combination of favorable circumstances to succeed. This deal in the contest between Canada and Germany required help from the defense and an end play.

Board 3
Dealer: South
East-West Vulnerable

NORTH
♠ Q J 5
♥ K Q 10 2
♦ J 7 4
♣ A 3 2

WEST
♠ 8 7 4
♥ A 5
♦ A 9 6 3 2
♣ Q 8 4

EAST
♠ K 3 2
♥ 9 8 4 3
♦ Q 10 8 5
♣ J 10

SOUTH
♠ A 10 9 6
♥ J 7 6
♦ K
♣ K 9 7 6 5

West	North	East	South
			1♠ *
Pass	2♣	Pass	3♣
Pass	3♠	Pass	3NT///

* Canape

A light opening by South (Nippgen) led to a forcing sequence and he settled into 3NT for lack of anything more appealing. Luckily this contract created value for his singleton ♦K when a low diamond was led at trick one.

Nippgen's play was astute. After winning the first trick with the singleton ♦K, he led a low heart. West ducked. This looked like a safe play but wasn't as West lost the opportunity to run the diamond suit. Nippgen then picked up the ♠K with a finesse. On the fourth spade, West was squeezed; if he discarded a diamond, declarer would lead hearts to his ♥A, giving up one heart and three diamonds to make the game. So instead West discarded a low club. Nippgen then ran five clubs, making 11 tricks and earning 6 IMPs for Germany.

North-South had their 24 high-card points and a qualifying suit. Even so there was some good fortune helping Nippgen bring home this close game. It was so close that it surely would not have been made if North had been declarer, thus exposing South's singleton ♦K in the dummy.

*　　*　　*　　*

It is easy to understand how a good five-card suit can produce an extra trick or two when partner has something of value in the same suit. But what about all of the times when you don't have much of a fit? Or to put it another way, what good is a qualifying five-card suit when partner has no support? While it may come as a surprise, quite often the two hands are compatible in some other way that compensates for the lack of tricks available from the qualifying suit.

> **When you cannot find tricks where you hope to find them, look elsewhere; you will be rewarded more often than your opponents think you deserve.**

*　　*　　*　　*

WORLD BRIDGE OLYMPIAD - 1996. This quarterfinal deal was between France and Chinese Taipei. The East-West hands fit the 05-03-24 combination exactly. Both teams declarer 3NT well prior to anyone having heard how this combination unlocks good notrump games.

Board 73 (Rotated)
West Dealer
North-South Vulnerable

NORTH
♠ K 10 5
♥ Q 10 6 2
♦ A K 8 6 3
♣ A

WEST
♠ Q 4
♥ K 7 4 3
♦ Q 4 2
♣ K 7 3 2

EAST
♠ J 9 7 3
♥ J 8
♦ J 10 7 5
♣ Q J 4

SOUTH
♠ A 8 6 2
♥ A 9 5
♦ 9
♣ 10 9 8 6 5

West	**North**	**East**	**South**
Taipei	***France***	***Taipei***	***France***
Pass	1♦	Pass	1♠
Pass	2♥	Pass	2NT
Pass	3♦	Pass	3NT///

In the open room, France sitting North-South bid every suit but clubs along the way to 3NT. North held a good five-card diamond suit opposite a singleton. The bidding was reported as "overbid slightly to reach a pushy but not unpromising 3NT". Today this would hardly be deemed pushy.

This is an infrequent deal where North's good suit is not supported by two or three in the other hand. Ponder it for a moment. There are just six tricks off the top. Neither diamonds nor clubs are promising. Can you get three more tricks in spades and hearts? This looks rather daunting and most of us would just as soon play fast and go on to the next deal.

West led the ♣2, the only unbid suit. The dummy ♣A won the first trick perforce, East playing the ♣Q to unblock the suit. Declarer (Szwarc) played the ♥2 to his ♥9 and West's ♥K. Ah, his prospects were looking up. West returned another club to partner's ♣J, who then switched to the ♥J. (Another club would have set up declarer's long clubs.) Of course it was too late. Szwarc won in hand with the ♥A and led a third club to establish the suit while holding the ♠A for entry: ten winners, plus 630.

* * * *

BERMUDA BOWL - 2007. The Championship came down to USA1 and Norway, Norway showed their mettle early, bidding these hands to 3NT while USA1 settled comfortably in 1NT.

Board 4
Dealer: West
Both Vulnerable

	NORTH	
	♠ K 9 7 4	
	♥ A Q 3	
	♦ A 8 2	
	♣ A 5 2	
WEST		**EAST**
♠ A 6		♠ Q 10 5 3 2
♥ 10 6 5 2		♥ J 8 4
♦ K 7 6		♦ Q J 9 5
♣ J 8 6 4		♣ Q
	SOUTH	
	♠ J 8	
	♥ K 9 7	
	♦ 10 4 3	
	♣ K 10 9 7 3	

West	**North**	**East**	**South**
Pass	1♣	Pass	1♦
Pass	1NT	Pass	3NT///

Norway's Tundal opened and partner pressed on to 3NT. They had 24 HCP and qualifying club suit. East led a low spade to partner's ♠A, who promptly returned a spade to Tundal's ♠K in hand. Starting on the good club suit, Tundal played the ♣A, fortunately collecting the ♣Q from East. The rest was easy. He gave up one club to West, who was unable to play to partner's good spades. Tundal made nine tricks for +600.

In the other room USA1 stopped at 1NT. On a diamond lead from East to West and a switch to a low spade from West, declarer (Katz) took seven tricks for his contract and +90, but it was an 11 IMP swing to Norway.

* * * *

In IMP contests there is little incentive to take overtricks as they are worth only one or two IMPs; and there is no percentage whatever if the effort risks the contract even by an iota. Consequently, it is not uncommon to see a declarer making game in one room while only part score in the other room, both however making their contracts.

It takes more than strict reliance on point count to make close calls at the table. Accurate notrump bidding requires you to take into account quality five-card suits. They are not niceties; they are essentials toward accurate bidding.

There is nothing new in recognizing shape in bidding. Years ago, Terence Reese, renowned player and author, wrote:

> **Players who count points and don't take note of distribution are a menace.**

The absence of a qualifying suit is equally significant, but as a contrarian indicator. With 24 high-card points (or less) and no qualifying suit, bidding game is not just pushy, it is moving into negative territory, and that is losing bridge.

13

ROAD TO GLORY

I am pleased that you have stayed with this notrump odyssey thus far. My belief and motivation is that the notrump strain is substantially under-appreciated. Surely this is the case with casual bridge players; and the evidence I see says it also is the case to a lesser extent among tournament players.

I suggested various bidding preferences that favor notrump over suit contracts. So long as your bidding framework is Standard American/Five-card Majors, these preferences are easily adopted as they involve calls that are readily available within the standard framework.

Notrump Openers

Marty Bergen's advice to open 1NT rather than a five-card major is right on. The bidding proceeds as it would following any 1NT opener. You are substituting one standard opener for another, and partner has all of the usual bidding tools available in response.

The underlying principle is that notrump contracts play as well or better whenever both hands are more or less balanced. Consequently the notrump option gets you to a higher-scoring contract more often than not, and more often that contract is notrump. These deals occur so often that you must prefer notrump to win match-point bridge consistently.

As dealer of the moment, I deal you this teaser.

♠ A K 9 7 5
♥ 7 4
♦ Q J 2
♣ A Q 8

You are dealt 5-3-3-2, a very good spade suit, and a worthless doubleton. Don't be a wimp; open 1NT, always.

Part Score Contracts

When partner opens 1NT and your potential is limited to part-score, it is better on average to play in responder's five-card major than in 1NT. The recommended practice is to transfer then pass (to play 2♥ or 2♠). Partner opens 1NT and you hold:

♠ J 9 7
♥ Q 7 5 4 3
♦ 9
♣ 8 6 5 2

The only value here is the heart suit as trump. Bid 2♦ (transfer) and then pass partner's 2♥ response. At least you will get a couple of heart tricks out of it.

When playing Stayman and Transfer conventions, you do not have the option to play in 2♣ and rarely in 2♦; so holding five clubs or five diamonds, pass to play 1NT.

Notrump vs. Minors

Playing in notrump rather than a minor is a preference that most everyone recognizes. There are many deals where you have reasonable choices between a minor and notrump. Suppose you hold:

WEST	EAST
♠ Q J	♠ 10 8 7
♥ A J 10	♥ Q 7 6 4
♦ K 7 6 5 3	♦ Q J 8 3
♣ K 7 6	♣ A Q

West	North	East	South
1♦	Pass	1♥	Pass
1NT	Pass	?	

Holding this East hand you see there is a 4-4 (or better) diamond fit. But so what? Notrump looks good and scores much better, so I hope you are not tempted to show diamond support next. This deal should be played in notrump and a diamond raise just gives information to the opposition. Even worse, partner may envision your hand as highly unbalanced and then bid . . .you name it. The only important question here is whether to pass or raise to 2NT.

Remember Hammon's rule? Perhaps it should be modified as follows:

> **When notrump is a choice among reasonable bids, take it, *especially when the alternative is a minor.***

There is a minor exception (pun intended). When partner opens 1NT and you have a six-card minor and only part-score potential, the deal will more often play better in the minor; in which case you may use a minor-suit transfer and play the contract in the minor at the three-level rather than 1NT.

WEST	EAST
♠ A J 5 3	♠ 9 8 7
♥ A J 10	♥ Q 7
♦ 10 2	♦ Q J 9 8 4 3
♣ K Q J 3	♣ 2

West	North	East	South
1NT	Pass	2♠	Pass
3♣	Pass	3♦	

Flats

One of the most important preferences involves flats. In fact it is so important that it rises to the level of "must".

Always steer the bidding to notrump when your hand is flat.

So that you do not get carried away with a four-card major, I suggest you mentally down-grade it to three cards so you are not tempted to go for the major (flat hands only.) As Marty Bergen said "There are absolutely no exceptions."

♠ Q J 10 9
♥ A 8 7
♦ K Q 6
♣ 8 6 4

Partner opens 1NT. You like these spades? Forget it; raise to 3NT.

♠ A Q J 10
♥ A 8 7
♦ K Q 6
♣ J 6 4

As you open 1NT, partner says 2♦ (transfer). Of course you call 2♥ as requested. If partner next calls 3♥, raise to four. Partner is in charge; do not second-guess him.

New Suits

You should ever be aware that every bid you make gives as much information to your opponents as to your partner. Of course you know this. But how much thought have you given to the consequences? Consider these three questions:

- Which side gains the most from a particular call?
- Do you really need to bid another suit to find the preferred contract?
- Should you mention a minor suit when notrump is a reasonable option?

There are many opportunities to streamline your bidding and keep the opponents in the shadows. These opportunities often involve minor suits and flat hands. When either hand is balanced, a minor suit game only rarely, almost never, outperforms notrump. It is even more futile to go for a minor suit game when one hand is flat.

WEST	**EAST**
♠ A 9 5	♠ K 8 7
♥ 7 6 5	♥ Q J 10
♦ A K 8 5	♦ Q J 10 9 3
♣ K 8 3	♣ Q 5

West	**North**	**East**	**South**
1♦	Pass	3♦	Pass
?			

West opens 1♦ with 14 high-card points. East must choose between 1NT and a healthy diamond raise. Here East decides 1NT would be an understatement, so she jump-raises diamonds. West may be tempted to try for a diamond game, but with this flat hand that would be foolish.

Qualifying Five-Card Suits

A five-card suit containing at least three honor points nearly always produces extra values in notrump contracts. This is as certain as any rule in bridge. Consequently, the odds favor making a notrump game with 24 high-card points and a qualifying suit. This is a central theme toward advancing your performance to a higher level. The reverse is equally valid: 3NT is a poor gamble with 24 honor points absent a qualifying suit.

*   *   *   *

Women's Team Olympiad - 2004. Here we find USA against Russia in the women's final. Halfway through the match, USA was barely ahead 137 to 131 IMPS.

Board 50
Dealer: East
North-South Vulnerable

NORTH
♠ K 6
♥ A 9 5
♦ K 9 7
♣ A K 10 9 7

WEST
♠ 10 3
♥ J 10 6 3
♦ 10 8 5 3
♣ Q 5 3

EAST
♠ Q J 8 4 2
♥ K Q 8
♦ A 6 4
♣ 6 4

SOUTH
♠ A 9 7 5
♥ 7 4 2
♦ Q J 2
♣ J 8 2

West	North	East	South
		1♠	Pass
1NT	Dble	Pass	Pass
2♠	3♣	Pass	3NT///

Apparently, West put great value on her three tens. Not to be cowed, North (Russia's Gromova) thought her hand was too strong for an overcall; and so she doubled and partner pushed on to game.

With South declarer, the lead came from West who started with the ♥J, a good choice because declarer (Ponomareva) had only one heart stopper. Ponomareva ducked; East won with the ♥Q and returned the ♥K and then a third heart, Ponomareva winning perforce with the dummy's ♥A.

Ponomareva cashed the ♣A then returned to her hand via her ♦Q. Now the club finesse was all that was needed to secure her 3NT contract (five clubs, one diamond, one heart and two spades).

In the other room USA held the North-South hands and settled for a part-score.

West	**North**	**East**	**South**
		1♠	Pass
Pass	1NT///		

This 1NT overcall by Molson (USA North) had a wide range of 10 to 17 points. It was passed out. East led the ♠Q, Molson winning with the ♠K in hand. She immediately began clubs from the top down, giving up the third club the West's ♣Q. A spade return gave Molson a third spade trick and nine tricks total.

The North hand was simply too good for an overcall. This deal gave Russia 10 IMPs to put them in the lead; and they went on to win the championship by a score of 271 to 259.

* * * *

Applying the combination 05-03-24 (remember: 5-card major - 3 honor points - 24 high-card points in the two hands) is more complex than other bidding preferences. The easiest configuration is when responder has the qualifying suit. For example, partner opens 1NT (15-17) and you hold:

♠ K J 3
♥ 7 5 4
♦ K Q 9 8 7
♣ 8 5

You have 9 high-card points added to partner's minimum of 15. You may confidently jump to 3NT on the strength of these diamonds. Alternatively you could remain conservative and invite to game with a 2NT raise (not recommended).

The situation differs when the strong hand holds the qualifying suit. Suppose you add two points for a qualifying suit and open 1NT with 13 to 15 high-card points. Partner responds normally, which is to assume you have a hand worth 15 to 17 points. With partner giving normal responses, you arrive at your desired notrump contract. So far, so good. The difficulty arises when, after you have valued your hand for notrump, partner directs the contract into a major suit game. The incremental value of a qualifying suit accrues in notrump but not in a suit contract. Hence you could find yourself in a rather thin major-suit game.

PARTNER	YOU	
	1NT	♠ A 6
2♦	2♥	♥ Q 8 2
4♥		♦ 10 9 8
		♣ A K 7 6 5

You have 13 high-card points. Since it is balanced and contains a qualifying suit, open 1NT. Suppose partner has a shapely hand with a good six-card heart suit. He values his hand to be worth 11 points and decides the best contract will be hearts. He calls 2♦ (transfer) then jumps to a heart game. All of this is standard of course, except that you have two honor-points less than partner expects. On the other hand, between the two hands there are nine hearts; and the ninth heart compensates nicely for the shortage of honors.

There is a side benefit to being aggressive which mitigates the risk; that is the preemptive value of notrump. Suppose left hand opponent (LHO) holds:

♠K9542 ♥5 ♦A96 ♣AQ43

If you open a minor, LHO will overcall 1♠. If you open 1♥, LHO has a good takeout double. But if you open 1NT, LHO has no good call.

Playing for Glory

Hopefully the bidding system, whatever yours may be, gets you into the best contracts most of the time. Beyond this, the great challenge comes in playing the hands. One small success at a time is the road to glory.

Sometimes you have the wherewithal to take just a couple of tricks; while at other times you may take what you justly deserve - most or all of the tricks. Whatever the potential, tournaments are won by taking one more trick than most of the others. Applying all of the knowledge available to you along with the positive and negative inferences - why didn't she lead diamonds? - will deliver you into the winners circle.

* * * *

SAN DIEGO FALL NABC - 2009. On the way to winning the Kaplan Blue Ribbon Pairs, Bramley/Fallenius stretched this deal to a notrump game with minimal honors.

NORTH
♠ A 9
♥ 10 9 7 4
♦ A K 4
♣ J 8 7 4

Dealer: East
Both Vulnerable

WEST
♠ K J 2
♥ K 6 5 2
♦ 10 9 2
♣ Q 10 3

EAST
♠ Q 10 6 3
♥ 8 3
♦ J 8 6 5 3
♣ A 6

SOUTH
♠ 8 7 5 4
♥ A Q J
♦ Q 7
♣ K 9 5 2

West	North	East	South
		Pass	1♣
Pass	1♥	Pass	1NT
Pass	3NT///		

North-South contracted for 3NT, with East-West passing throughout. West led the ♦10. Declarer Bromley in the South seat could see he would need some luck or aid to pull in nine tricks. There were seven in sight, counting three hearts, three diamonds, and the ♠A. And there was the 50/50 possibility of a fourth heart. Otherwise he would need two clubs.

He took the first trick with the ♦K, hiding his ♦Q from the defense. Next he played a heart to the ♥Q and West's ♥K. West continued diamonds, hoping the dummy ♦A was declarer's last diamond stopper. Not so; Bromley took this trick in hand with the ♦Q then boldly led the ♣K. East won with his ♣A and continued to attack diamonds even though he had no more entries. Bromley won with the ♦A, took two top hearts in hand, and led a club toward dummy. With the ♣Q onside, nine tricks were now in the bag.

This game produced 34 matchpoints out of 38 maximum. The defenders would not give up on diamonds once they were committed at the opening lead.

*   *   *   *

In theory, you should be playing offense half of the time and defense the other half. However, if that is the way it turns out, then you are not winning. You must play offense more than half of the time, thus keeping the opponents in shadows and the tricks coming your way. When on offense, you have opportunities to win unearned tricks far more often than you have on defense.

We have seen how some of the world's top bridge players played, and mostly made, difficult notrump contracts. They played offense in ways that increased their odds from 50/50 or so upwards towards 100%. They employed a variety of mental gymnastics - assimilating all sorts of information gleaned from opponents bidding or not, choices of opening leads; and above all keeping track of a variety of data bytes such as empty spaces, honors held or denied, and more, some beyond codification.

*   *   *   *

VENICE CUP - 1991. The championship contest between Austria and USAII had only 14 boards to play and the Americans were well ahead; the Austrians coming on strong.

Board 115 (Rotated)
Dealer North
East-West Vulnerable

NORTH
♠ Q 8 3 2
♥ 9 8 5
♦ 10 8
♣ A 10 8 2

WEST
♠ J 9 5
♥ Q J 7 4
♦ K 9 7 6 2
♣ 5

EAST
♠ A 10 4
♥ K 10 3
♦ J 4 3
♣ 9 7 4 3

SOUTH
♠ K 7 6
♥ A 6 2
♦ A Q 5
♣ K Q J 6

West	North	East	South
	Pass	Pass	1♣
Pass	1♠	Pass	2NT
Pass	3NT///		

Austria (North-South) contracted for 3NT and declarer was South (Erhart). Erhart won the diamond lead from West in hand with the ♦Q. She had eight tricks in sight and needed to find a second spade trick.

Erhart immediately played four rounds of clubs, West discarding heart, a spade, and a diamond. Next Erhart played the ♠7 to the ♠Q and East's ♠A; she won the diamond return with her ace.

The six-card end position was:

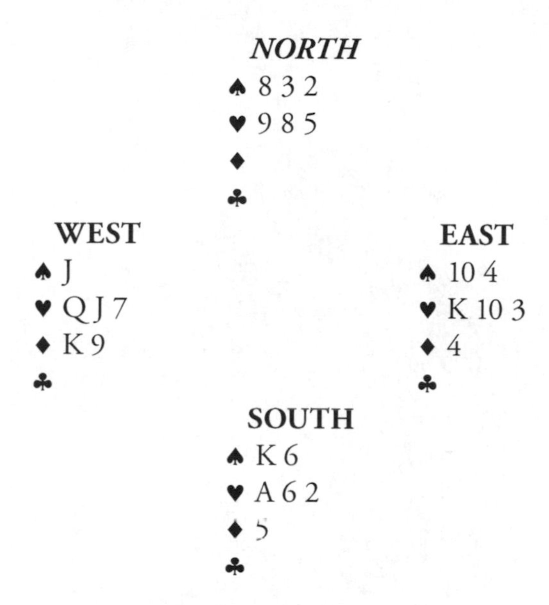

To this point, Erhart had won four clubs and two diamonds, while the defenders had only one, the ♠A.

Erhart led her last diamond to West, discarding a spade from dummy. When West then played her 13th diamond, Erhart discarded another spade from dummy, and East also discarded a spade. At this stage, North, East and West were down to a single spade so Erhart won the next trick with the ♥A, and her game-winning ninth trick with the ♠6.

GLORY BE

A. INDEX OF TOURNAMENT DEALS

1999 Bermuda Bowl	Round Robin: Bd 13	France-Norway	96
1999 Venice Cup	Bd 16	France	43
1999 Venice Cup	Round Robin: Bd 14	Great Britain-Germany	42
2000 Venice Cup	Round Robin: Bd 20	Great Britain-China	7
2004 Women's Team Olympiad	Final: Bd 50	USA-Russia	132
2004 World Team Olympiad	Final: Bd 25	USA-Russia	77
2004 World Team Olympiad	Final: Bd 137	Italy-Netherlands	70
2004 World Team Olympiad	Round Robin: Bd 18	Netherlands-HongCong	76
2007 Bermuda Bowl	Final: Bd 4	USA1-Norway	125
2007 Bermuda Bowl	Final: Bd 77	Norway-USA1	71
2007 Bermuda Bowl	Semifinal: Bd 55	Netherlands-Norway	27
2007 Senior Bowl	Bd 22	Argentina-Indonesia	16
2007 Venice Cup	Quarterfinal: Bd 81	USA2-France	99
2007 Senior Bowl	Semifinal: Bd 8	USA1-USA2	6
2007 Senior Bowl	Round Robin: Bd 68	Italy	67
2007 Senior Bowl	Round Robin: Bd 2	USA1-Canada	72
2007 Senior Bowl	Bd 87	USA1-France	5
2009 Fall NABC	Pairs	US Pairs	135

B. INDEX OF FEATURED PLAYERS

C. TEAMS OF FOUR

Suppose you invite seven friends to play bridge, and you divide into two teams to play team duplicate. Each team organizes itself into two pairs – so you have team A (pair 1 and 2) and team B (pair 3 and 4). Each pair of a team is assigned to one of two rooms to play against a pair of the other team. It is important to play in different rooms because both tables will be playing the exact same deals. Like a pairs duplicate game, decks are shuffled, dealt and placed into boards that will be played in one room and passed on to be played in the other room. The key to team contests is that one pair of each team sits North-South in one room and East-West in the other room, playing the exact same deals. The consequence of this arrangement is that each team plays every board both ways. This is the essence of team duplicate.

Suppose for board #1 (non vulnerable). team A playing E-W, contracts for 3 spades and makes the contract for a score of +140. In the other room where team B plays E-W, they bid and make 3NT, for a score of +400. The difference is 260 points in favor of team B.

You could score the contest in total points, in which team B would have earned 260 points more than team A. Alternatively you could simply count each board as a win or loss, in which case team B would be ahead 1 to 0. Major team tournaments are usually scored in International Matchpoints, or IMPs. To do this, you would look up the IMP value of 260 points to determine how many IMPs team B earned on the deal. These 260 points convert into a gain of six IMPs.

There is an official chart that translates contract-score differences into IMPs. You can find it in the ACBL's Laws of Duplicate Contract Bridge. Some significant features of IMP scoring are:

- A major suit game earns 1 IMP more than a 3NT game.

- Making an overtrick when the opposition just makes the same contract earns you I IMP more.

- When you double and set a 3NT contract by one trick, the double is worth 3 IMPs to your team. Alternatively, if the opponents make the doubled contract, it cost your side 3 IMPs.

- The 500 point premium for a small slam (not vulnerable) is worth 11 IMPs.

- Most large scores come from one team bidding and making game while the other is set or does not bid the game.